MW00967082

Contents

THE NEW ROUTER HANDBOOK

Patrick Spielman

 Sterling Publishing Co., Inc. New York

Library of Congress Cataloging-in-Publication Data

Spielman, Patrick E.
      The new router handbook / by Patrick Spielman.
            p.   cm.
      Includes index.
      ISBN 0-8069-0518-2
      1. Routers (Tools)   I. Title.
TT203.5.S624   1993
684'.083—dc20   93-25637
                                           CIP

Published by Sterling Publishing Company, Inc.
387 Park Avenue South, New York, N.Y. 10016
© 1993 by Patrick Spielman
Distributed in Canada by Sterling Publishing
% Canadian Manda Group, P.O. Box 920, Station U
Toronto, Ontario, Canada M8Z 5P9
Distributed in Great Britain and Europe by Cassell PLC
Villiers House, 41/47 Strand, London WC2N 5JE, England
Distributed in Australia by Capricorn Link Ltd.
P.O. Box 665, Lane Cove, NSW 2066
Manufactured in the United States of America
All rights reserved

Sterling ISBN 0-8069-0518-2

6

Acknowledgments

A book of this scope is impossible to assemble without the generous assistance of many experts and specialists. I've been extremely fortunate to meet, visit, and correspond with numerous manufacturing engineers, product managers, woodworking craftsmen, and other terrific individuals from around the world who provided important sources of information. I am extremely grateful to all who have afforded me their time and effort to help make this an informative and helpful volume.

First, thanks must go to my good friend Charlie, for providing the idea, the motivation, and the "pressure." And, thanks to my editor, Mike Cea, who always does the impossible within rigid time constraints. Much gratitude and love is expressed to my wife, Patricia, for keeping our businesses on course and affording me the freedom to devote my time and energy to this effort. I sincerely thank my son, Bob, of Spielman's Cedar Works, for being a "sounding board," jig tester, and personal consultant.

Carlo Venditto, Marcello Tomassini, and Cliff Paddock of CMT Tools and their staff generously provided much practical and innovative technical material concerning router bits, for which I am extremely grateful. I also thank and recognize R.S. and Mark O'Brien of Onsrud Cutter, and Rob Van Nieuwenhuizen of Furnima Industrial Carbide, for their help. Chris Carlson of Bosch, Jim Brewer of Freud, Jim Phillips of Trend, Glen Davidson of Welliver and Sons, and Dennis Huntsman of Porter-Cable all were exceptionally helpful with material about their products. Thanks to Dick Jarmon for his input on router history, to Chris Taylor for his help and the great cover photograph, and to Roger and Tom McIlree for the outstanding alphabets. Thank you, woodworker Curt Whittington, engineer Mark Obernberger, and Mr. Obernberger's crew at Palmer-Johnson, Inc., for the special tips and ideas.

Thanks to my artistic daughters, Sherri Valitchka and Sandra Fridenfels, for improving my photos. Thanks to ProGrafx and Dirk Boelman for the drawings, and a very special note of thanks goes to my typist, Julie Kiehnau, for effectively following the "arrows" and promptly turning all my rough scribbles into finished copy.

I also thank the following individuals and companies for photos and technical assistance:

Advanced Machinery Imports
AEG Power Tool Corp.
Align-Rite Tool Co.
Amana Tool Corp.
Beall Tool Co.
Black & Decker
Robert Bosch Power Tool Corp.
Brad Park Industries, Inc.
Richard Byrom
Cabot Safety Corp.
Carbide Alloys, Inc.
Cascade Tools
Constantine
Craftsman Tools
Custom Wood Mfg.
Delta International Mach. Co.
DeWalt
Dremel Power Tools
Dupli-Carver
Eagle America Corp.
Edge Finisher Corp.
Ekstrom, Carlson & Co.
Elu
Excalibur Machine & Tool Co.
Formica Corp.
Fox Mechanical Products
Freud
Furnima Industrial Carbide, Inc.
Walter Hartlauer
Hartville True Value
Hitachi Power Tools, USA Ltd.
Hoffmann Machine Co.
Dick Jarmon
JDS Company
JCM Industries
JoinTECH
Kehoe Mfg. Co.
Keller & Co.
Kimball Machine Co.

Leigh Industries
Lewin Router Compass
Makita
Metabo Electric Power Tools
MLCS Ltd.
Mark & Cut Co.
Marlin Industries Inc.
Mule Cabinet Machine, Inc.
NuCraft Tools, Inc.
Olive Knot Products
Onsrud Cutter
CMT Tools
C.R. Onsrud, Inc.
Paso Robles Carbide
Phantom Engineering, Inc.
The Pinske Edge
Porta-Nails, Inc.
Porter-Cable
Quality VaKuum Products
Racal Health & Safety, Inc.
R.B. Industries
Reliable Cutting Tools
Rig-A-Mortise
Carl Roehl
Rousseau Company
RYOBI America Corp.

S J Fine Woodworks
Sandarc Industries
Sears
Shopsmith
Skil Corp.
Soss Hinges (Universal Industrial Products)
Stanfield Mfg. Co.
Superior Moulding Co.
Taylor Design Group, Inc.
Titman Tip Tools, Ltd.
Trend Routing Technology
Trim Tramp Ltd.
Unique Machine & Tool Co.
Vacuum Pressing Systems
Vermont America Corp.
Ed Warner
Pat Warner
Welliver & Sons
Curtis Whittington
Wisconsin Knife
Witmer Wood Products
Woodhaven
Wood Miser Products
Woodrat
Woodsmith magazine
Woodworkers Supply, Inc.

Introduction

The router was first introduced around the turn of the century, and soon became recognized as woodworking's most versatile and useful power tool. Today, because of improvements in the routers themselves and in the products used in the routing process, these tools are even more versatile, and remain far and away the most popular power tool.

There are many reasons why the router is so popular. It is ideal for making perfect joints for all kinds of furniture and cabinetry, including kitchen and bath cabinetry. Even normally difficult joints such as mortise-and-tenon, box, and dovetail joints of all kinds can be easily made with the router (Illus. i-1). You will also find it invaluable when building kitchen and bath cabinetry. There are special cutters available with which you can create raised panel doors (Illus. i-2), cut and trim plastic laminate countertops, and assist in all of the carcass joinery. Distinctive architectural panels (Illus. i-3), doors, windows, and other special millwork for your home can all be fabricated from router-cut components. The router also makes perfect raised or incised letters and designs for signs.

Illus. i-2. Cabinet doors with framed raised panels produced by the Superior Moulding Co., Troy, Alabama.

i-1. Dovetails of all types including the decorative double dovetail shown here are easy to make with the router and appropriate bits. Refer to Chapter 26 for information about router dovetailing. (Photo courtesy of Taylor Design Group, Inc.)

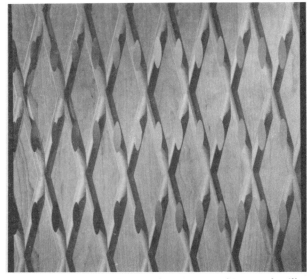

Illus. i-3. These beautiful and decorative architectural grilles are router-cut.

With special accessories, the router can do unusual tasks that include duplicating wood carvings, making and threading wood dowels, and cutting plain or decorative round turnings (Illus. i-4).

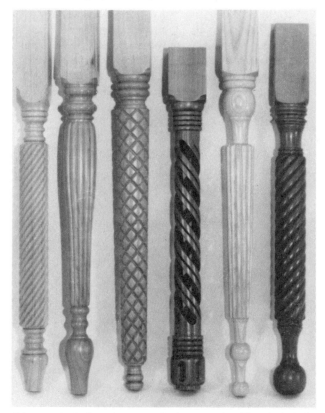

Illus. i-4. Think these legs were lathe-turned and hand-carved? They were cut with a router in a device called the Sears Router Crafter. It's described in Chapter 33.

The router is not limited to cutting wood. It is widely used in industry to cut plastics of all kinds and soft metals, including aluminum, brass and copper. I know of a company that cuts letters from thick aluminum, and another company that cuts out and bevels ½″ aluminum parts with the router for welding, and then smooths the welded joints with the router.

The router can be used freehand to cut irregular designs into or through a variety of materials (Illus. i-8). You can also guide the router with templates and patterns to cut duplicate designs of any conceivable profile, or to make straight-line cuts so smooth and true they compare to the edge work performed with large jointers.

Part of the reason for the router's versatility is that it

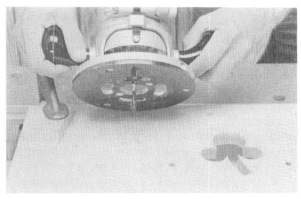

Illus. i-5. This pierced cutout normally requiring a drill and a scroll or sabre saw was cut entirely with the router.

can be held in your hand or fixed in place with the work taken to the router. This book examines both methods of using the router.

In addition to using the router in the normal upright (vertical) position, it is also commonly used inverted and sideways (horizontally) in router tables of various designs (Illus. i-6). Chapters 20–24 explore router tables and techniques for using them, because this is one of the most popular uses for the router today.

In the following pages, information is also presented on how to make and use patterns, templates, and jigs with which to exploit the full potential of the router. For example, some of the aids described will enable you to

Illus. i-6. This router table, designed and fabricated by the author, features an inverted, under-the-table router and an adjustable plate to accommodate the router in a horizontal position.

make identical cuts or produce shapes in quantities of 2 to 2,000 or more, as your needs require (Illus. i-7). Thus, you will be able to operate your router inexpensively and efficiently in a minimum of space.

The router is the first tool I turn to when confronted with a wood cutting or shaping problem. For example, I have always been tempted to try hand-carving wood chain, which is visually quite appealing, but was put off by the amount of tedious work that would be involved. Using a router, I was able to make a chain that, though not authentically hand-carved, is made of wood and has perfect links. Chapter 35 explains how this is done.

This book includes guidelines for buying both routers and bits, so you can avoid making the same expensive mistakes I've made over the years. Many different brands of router are examined and evaluated (i-9). Information is also provided on how to choose from the hundreds of router bits available (i-10). I have also included descriptions of a variety of router accessories and related router machinery from around the world.

In the following pages, I've tried to answer any questions and concerns about routing you may have. However, it is simply not possible to write a book on routers that is totally inclusive because, as a friend and router expert in England, says, "The subject of the router is simply inexhaustible." Still, you will find that the information presented here will help you use your router to its greatest potential.

Patrick Spielman

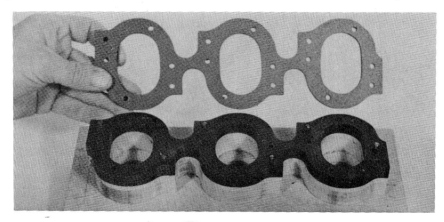

Illus. i-7. Pattern routing to duplicate cuts and parts precisely. Here ¼″ plastic pieces are triple-stacked and all cut together in just one pass.

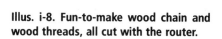

Illus. i-8. Fun-to-make wood chain and wood threads, all cut with the router.

Illus. i-9. A sampling of just some of the routers available today.

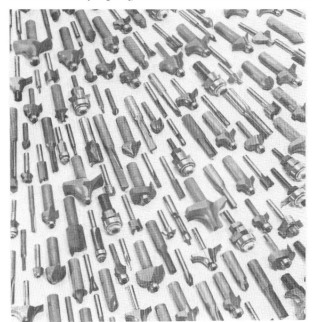

Illus. i-10. A few of the types of bits available.

Basic Information

1
Historical Overview

The "router" that we are familiar with today probably evolved from a specific type of early tool called the router plane (Illus. 1-1 and 1-2). Tools of this type were fitted with cutters to scoop out V-shaped, round, and flat-bottom grooves in wood surfaces. Their open-base design looks similar to that of fairly recent routers.

The first devices to power rotary cutters were foot-powered. They were actually the forerunners to our present-day router and shaper tables (Illus. 1-3). The Barnes Foot Power Former was manufactured by the W. F. & John Barnes Co., of Rockford, Illinois, which was established in 1872. Machines of this type were used throughout the turn of the century. Many foot-powered tools were used well into the 1930s, especially in rural areas, which did not have electricity yet. Foot power was used to spin cutters for forming decorative edges on brackets, scrollwork, and panels, and to make regular and irregular mouldings of all styles on stock up to 7⁄8″ thick.

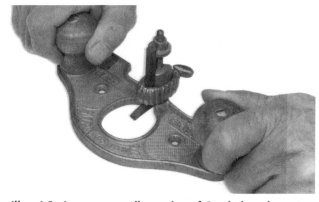

Illus. 1-2. A more versatile version of Stanley's early router-plane. Interchangeable cutters of different shapes could be mounted on the vertically adjustable post. Cutters could alternatively be inserted facing the rear so cuts could be made in tight areas, a feature definitely lacking in our modern electric routers.

Illus. 1-3. This Barnes Foot Power Former is the forerunner to modern router and shaper tables. This machine featured a patented velocipede foot power arrangement, attained speeds of up to 2,500 rpm, and had knives that could be rotated in either direction at the will of the operator and according to the grain of the wood.

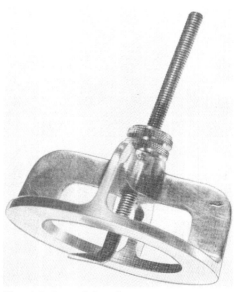

Illus. 1-1. An early hand router.

Spindle speeds of 2,000 to 2,500 rpm (revolutions per minute) was attainable. These devices featured one advantage not embodied in today's hand-held electric routers: the cutter knives could be rotated in either direction to take advantage of the grain of the wood. The table on the Barnes machine was adjustable vertically to 2¼″ below the cutting edges.

In the first edition of *Router Handbook*, I gave unsubstantiated credit to R. L. Carter, a New York pattern maker, for the development of the first portable electric standard or fixed-based router. However, there now appears to be evidence that the first portable electric router produced commercially was actually the Kelley router, first sold in 1905 (Illus. 1-4). Described as "a universal woodworking machine," the Kelley portable router was very large and heavy by present-day standards. Kelley routers were manufactured in Buffalo, New York, by the Kelley Electric Machine Co.

The Carter electric router, also developed in New York at Syracuse, was first sold during World War I (1914–1918). Carter routers were eventually manufactured in a wide variety of sizes ranging from ⅐ to 1½ hp and with weights of 3½ to 35 pounds. They were called "electric hand shapers." Carter manufactured over 100,000 in a ten-year span, and these routers earned the designation "the wonder tool" (Illus. 1-5).

The original Carter routers were low-performance tools, but a 1928 catalogue published by the R. L. Carter Co., Phoenix, New York, featured a number of Carter electric routers "for lamp socket operation." The catalogue also featured an intriguing selection of accessory devices for circle-cutting, beading and fluting, inlaying, and dovetailing, and many bits with designs similar to those available today. The Carter routers featured a motor housing and base frame with 16 threads per inch. A single full revolution of the motor within the frame changed the depth of the cut by ¹⁄₁₆″.

Carter sold his business to Stanley Electric Tools in 1929. Stanley continued to manufacture routers (Illus. 1-6 and 1-7) until the early 1980s, when the Stanley

Illus. 1-4. **Above:** The Kelley electric router was first sold in 1905. **Right:** Early routing in a factory producing large casks and wood tanks with a Kelley router. Note the makeshift compass control and sawhorse work supports.

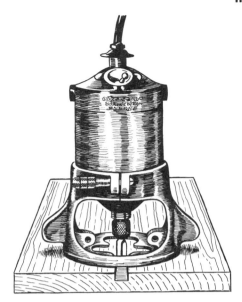

Illus. 1-5. This ½ hp Carter electric router appeared in the 1928 **Carter Products** catalogue of Phoenix, New York. This router took only ¼" diameter shank bits, featured a running speed of 18,000 rpm "without vibration from any light socket," weighed only 3½ pounds, and had a list price of $46.50. Routers were relatively expensive during this time.

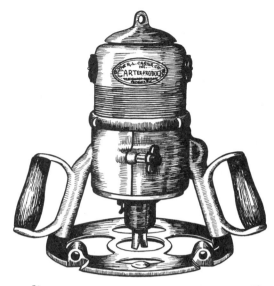

Illus. 1-6. The Carter heavy-duty router as it appeared in a 1928 catalogue. It had the following features: ½ hp, 12,000 rpm, a 11" diameter base, a threaded screw-in motor casing for accurate depth adjustment, a convenient switch, and a template guide that could take ¼" and ½" diameter shank bits. It weighed 35 lbs. and was listed at a selling price of $245 (with some bits and template guides).

Power Tools Division was purchased by the Bosch Power Tool Corporation.

The Porter-Cable Company founded in Syracuse, New York, in 1906 by R. E. Porter, G. G. Porter and F. E. Cable also produced a line of routers and provided some competition to the Stanley fixed-base routers until 1960, when Porter-Cable was purchased by Rockwell Manu-

facturing Co. In October of 1981, Rockwell divested itself of its portable power tool division and sold it to Pentair Inc., of which Porter-Cable was a manufacturing division. This returned Porter-Cable to the router manufacturing business.

Porter-Cable began production of routers in 1940. The company's current line evolved from one invented by

Illus. 1-7. The progressive changes in router design and manufacturing are evidenced in these old Stanley routers, the forerunners of many of the Bosch fixed-base routers.

Elmer P. Sacre, an employee of the Unit Electric Tool Co. of Syracuse, New York. Unit Electric Tool Co. began producing this router in 1946. The first type of router produced was a B model for a ¼″ shank bit, followed by a C model for a ½″ shank bit. These routers were sold with a full line of accessories, including a plane attachment, door lock mortiser, lock face template, and a stair template.

These forerunners of the Porter-Cable routers (Illus. 1-8) were sold under the label of UNIQUIP to convey the idea of one motor for all equipment. Unit Electric Tool Co. also produced floor- and bench-mounted router tables with tilting fences. Unit Electric was purchased by Porter-Cable in 1948, and in 1950 all routers were labelled Porter-Cable Speedmatics. In 1948, the B router sold for $125 and the C for $198, certainly not a cheap investment for a carpenter. In 1952, Porter-Cable began providing aluminum housings and the 8½-pound B model became the Porter-Cable model 100. In 1957, this tool was redesigned to ⅞ hp and less than 7 pounds. It is still in production today. The 18-pound C model became one of Porter-Cable's 1½ hp heavy-duty routers, weighing less than 10 pounds.

Plunge-type routers were first introduced in 1949 in Germany by Elu, a company named by its founder, Eugen Lutz, in 1928 (Illus. 1-9). Worldwide distribution of Elu plunge routers began in 1951. These routers quickly evolved to become perhaps the best known and most widely used plunge routers around the world. The original Elu plunge router underwent some design changes in 1960 but, except for its new electronic system, looks much the same today (Illus. 1-10). Elu routers are now made in Switzerland and distributed in the United States by the Black & Decker Company, which also manufactures routers.

There are many portable electric router manufacturers today. A list of United States manufacturers would include Sears, Dremel, DeWalt, Milwaukee, and Skil. A list of imports would include Freud (produced in Spain), RYOBI, Hitachi and Makita (produced in Japan), AEG (manufactured in Germany, but owned by Atlas Copco Electric Tool Inc. of Sweden) and Metabo (also of Germany). Taiwanese and/or other imports are available under the labels of Jepson Power Tools (Ko Shin Electric & Machinery Co., Ltd.) and Chicago Electric Tools.

Illus. 1-8. These routers are the forerunners of the Porter-Cable fixed-base routers. On the right is the UNIQUIP C model router. On the left is an early Speedmatic UB model, made in Syracuse, New York.

Another well-known portable power tool company that has some involvement in the router business is the Skil Corporation. In 1954, Skil began production of an all-aluminum-construction, ¾ and 1¾ hp fixed-base router. It discontinued production of this router in 1979. Between 1964 and 1974, Skil also produced a laminate trimmer. In 1987, Skil introduced a consumer 1¾ hp plunge router, and in 1991 a 2¼ hp production router.

Illus. 1-9. Early German-made Elu routers. Right: a 1930 fixed-base router. Left: the world's first commercial plunge router, produced in 1949.

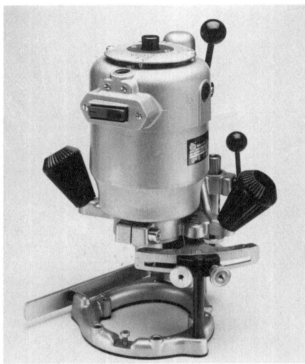

Illus. 1-10. A current version of an Elu router looks much the same as the first one sold in 1951. Not available in the United States, this router has some unusual adjustments not found on other Elu or any other plunge routers.

Routing Machines

The basic routing concept—a cutter mounted to the shaft of a high-speed motor—was applied to large, stationary, factory-type machines in the early 1900s. Today, there are very sophisticated high-production routing machines, such as CNC routers with hydraulic and air-feed systems and automatic computer-programmed controls, that cost from $50,000 to $500,000.

New, smaller versions of CNC routers in various sizes and cutting capacities are now available for the small production shop. You control the cutting path of these routers with your personal computer. One such unit is manufactured by Digital Tool, Inc., Pittsburgh, Pennsylvania (Illus. 1-11). This system costs approximately $10,000. It consists of a bridge with one or two programmable spindles. The customer has the option of supplying his own 3¼ hp motor unit or units or fitting the bridge with heavy-duty 5 or 7½ hp Perseke routing heads.

The customer can build his own 8′ × 12′ table to keep the initial costs down. The machine can be configured with two-dimensional-cutting capabilities or provided with a 6″ Z axis for doing three-dimensional work (carvings). Other bridges such as boring heads can be added to extend the automated machining capabilities. The manufacturer predicts that similar devices in scaled-down versions and in the $1,000–$2,000 range will be available in the not too distant future. They will permit the home hobbyist woodworker to link his personal computer to his router. This will probably greatly expand the capabilities of the router.

Illus. 1-11. Digital Tool's CNC router consists of one or two 3¼ hp routers on this 400-pound programmable bridge-and-track system powered or controlled by your personal computer. The customer has the option of building his own table and fixtures.

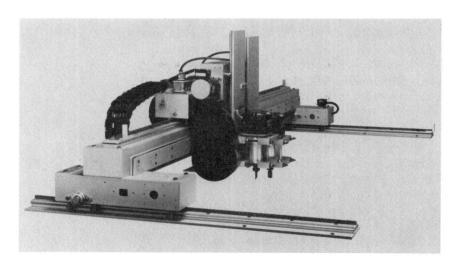

2
Types of Routers

There are two basic kinds of router: fixed-base and plunge (Illus. 2-1). This chapter examines the essential features these routers have in common.

Illus. 2-1. At left: a fixed-base, single-speed Black & Decker 1½ hp router. At right: a Hitachi 1½ hp, variable-speed plunge router.

Fixed-base or standard routers (as they are sometimes called) have bases that clamp directly to a removable motor (Illus. 2-2 and 2-3), making the router one integral or "fixed" unit. Plunge routers have motors attached to spring-loaded bases. The springs are located in or outside vertical poles or posts; this facilitates an independent vertical movement of the motor unit without the router user having to move or lift the router base from the work (Illus. 2-4 and 2-5). The motors of plunge routers are normally only detached from their bases for repairs or servicing.

Currently, fixed-base and plunge routers are the only two types of routers available. Each type has certain advantages and limitations over the other. Router design has been greatly improved in recent years, yet there is still room for considerable improvement. Routers are still much too loud. The sounds they emit frighten beginners and distress serious router users. Also, there is the problem of dust collection. A few manufacturers have made an attempt to incorporate some sort of apparatus for collect-

ing or removing dust, but the dust-collection systems used are generally ineffective and cumbersome.

Neither the fixed-base nor the plunge router is well suited to edge-routing and edge-forming—two operations the router is most commonly used for. The design of all routers is such that very little of the router's base is on

Illus. 2-2. Virtually all fixed base routers have removable motors.

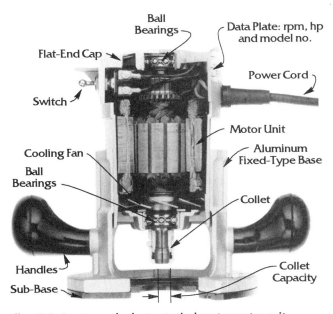

Illus. 2-3. A cutaway look at a typical router motor unit.

Illus. 2-4. Note the springs inside the tubular posts on the base of this Elu plunge router.

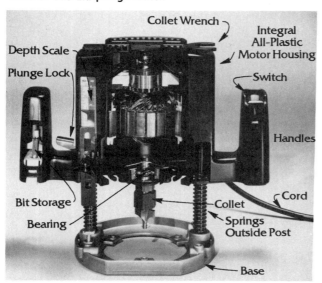

Illus. 2-5. A cutaway view of a Skil model 1835 plunge router with exposed springs.

the work during edge-routing operations (Illus. 2-6 and 2-7).

It is highly doubtful that there will ever be one ideal router designed to do everything well. Routers are continually used in new applications such as table-mounted work, which requires the router to function like a wood shaper/moulder. Fortunately, now router speeds can be controlled to match various sizes of cutters and different materials.

The best router to have depends entirely on the range and type of woodworking jobs it will be required to do. A very small, lightweight mini-router, such as the Dremel router shown in Illus. 2-8, is perfect for models, miniatures, and very light woodcrafts. A big 3¼ hp production router (Illus. 2-9) is obviously not suited for this type of work. It might be ideal for handling a variety of demanding cutting jobs in cabinet, furniture, and heavy-wood-product fabrication.

The majority of woodworkers needs a router that performs a broad range of tasks somewhere between the extremes described above. Determining which router is

Illus. 2-6. All current routers have the same inherent design limitations for edge-routing. The router's center of balance is not self-supported over the work, allowing the router to tip over.

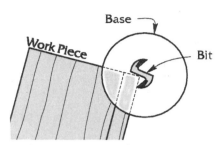

Illus. 2-7. Note how little of the router's base (less than a quarter of it) is supported on the workpiece when you are routing an edge around a corner, as shown here.

Illus. 2-8. This small, lightweight Dremel Moto Tool, rotating at 30,000 rpm and fitted to a router base with an edge guide, is being used to cut a narrow, shallow groove.

Illus. 2-9. Here a big, 3¼ hp, multiple-speed (10,000 to 21,000 rpm) plunge router makes deep surface cuts as it follows along a template pattern. Note the operator's eye and ear protection.

best suited for you is not a simple task. The most common mistake is to purchase a router entirely because of price. In a relatively short time, important parts of cheap models wear out or simply do not function as expected. The cheap model may not have the necessary horsepower and other conveniences that provide long-term routing safety and pleasure, thus undermining the reason for using the router entirely.

Serious amateurs and professional woodworkers eventually purchase a second and even a third router. Each purchase should be carefully analyzed, and the router selected for a particular kind of work.

There are many excellent general-purpose routers to choose from. Currently, there are more than 100 portable electric routing power tools available. These include related tools such as laminate trimmers and small hand carvers. You probably will select a router from those illustrated or described in this book. Comparative specifications of fixed-base and plunge routers currently available are given in Chapters 3 and 4. However, before you begin this selection, you should be familiar with the basic parts of the router, how these parts function, and other router features. This information is given below.

Motor Unit and Base

The two essential parts of every router are the motor unit and the base. Routers are essentially sized and designated according to the horsepower (or amps) of their motors. Most routers sold today have motors that range in horsepower from ⅝ to 1½ (Illus. 2-10 and 2-11). Those under one hp are for very light woodcrafting, those from 1½ to 2¼ hp are for general woodworking, and those from 3½ to 5 hp are generally strictly production tools intended for heavy cutting and continuous use (Illus. 2-12–2-14). Very small motors such as the .6 amp motor on the Dremel router turn small bits with shanks just ⅛"

Illus. 2-10. Shown here are light-duty, non-production routers by Black & Decker and Sears. These routers have ¼-inch collets and range in horsepower from ⅝ to 1½.

Illus. 2-11. Three light-duty imported routers with ¼" collets. On the left is a ¾ hp RYOBI. In the middle is a ¾ hp Jepson. On the right is a 1 hp Makita. Note the handle designs, their locations, and the hole openings in the router bases.

Illus. 2-12. Three durable, medium-duty, fixed-base routers. From left to right they are: a ⅞ hp Porter-Cable router, a 1½ hp Black & Decker router, and a 1 hp Milwaukee router.

Illus. 2-13. A selection of medium-duty plunge routers. Shown from top left to right are: a 1 hp AEG, a ⅞ hp Makita and a 1 hp Elu; shown from bottom left to right are: a 1¾ hp Skil, a 1 hp RYOBI, and 1½ hp Hitachi.

Illus. 2-14. Some of the big, powerful production routers. On the top row, from left to right, are a Hitachi 3½ hp V-speed router and a Bosch 3½ hp electronic variable-speed plunge router. On the bottom row, from left to right, are a Porter-Cable 3¼ hp 5-speed, fixed-base router, a RYOBI 3 hp variable-speed plunge router and Elu's big 2¼ hp variable-speed plunge router.

in diameter. Most other light-duty routers are more powerful and carry medium-size bits with ¼″ shanks. Large routers carry cutters with ½″ diameter shanks.

Motor Horsepower and Speed

Horsepower, more than motor speed, is required for making deep or heavy cuts in difficult-to-machine materials. The motor speed, referred to as revolutions per minute (rpm), of routers can vary widely. Single-speed router motors can range from 16,000 to 30,000 rpm.

High-motor speeds are not synonymous with increased power. In fact, the reverse is true. Usually the lower the horsepower, the greater the rpm. Low-horsepower, high-rpm motors will not hold up when cutting deeply into difficult hardwoods. This is not to say that lightweight routers are not good. Lightweight routers are simply designed for jobs involving shallow cuts, trimming, etc. They should not be forced to exceed their design limitations.

Generally, it's safe to assume that all routers have sufficient motor speeds for the work they are designed to do. Routers have much higher speeds overall than other types of woodworking tools. A hand drill seldom exceeds 2,500 rpm. Circular saws seldom rotate faster than 5,500 rpm. It is the high speed of the router that produces surfaces cut so smoothly that only minimal sanding, if any, is required.

Higher speeds, in theory, do produce smoother cuts. If you have a router operating at 20,000 rpm and one of equal horsepower rotating at 25,000 rpm, you will get 25 percent more cuts per inch or foot of cut surface from the faster model. A lightweight, high-speed, low-horsepower router can do many of the same jobs larger, greater-horsepower models do, but in limited capacities. For example, instead of making a deep cut in one pass, you will have to make the cut in two, three, or more successive passes at shallower depths. This practice is essential as a preventative measure, to minimize the possibility of excessive strain on the motor and bearings. This sort of inconvenience taxes one's patience.

Better routers now have variable- or multiple-speed electronic motors that put appropriate speed selection literally at your fingertips. Illus. 2-14 shows some of the bigger routers with electronic variable-speed controls. Most speed changes are made by rotating a dial on the router motor (Illus. 2-15–2-20). By shifting a small button/lever on the Porter-Cable router (Illus. 2-16), you can select one of the motor's five speeds. One of the most conveniently located is the speed selector dial on the

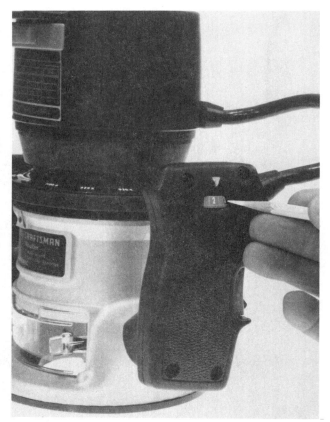

Illus. 2-15. This Sears 1½ hp variable-speed, fixed-base router features multiple speeds from 15,000 to 25,000 rpm (not much of a range) controlled by a trigger switch on the pistol-style grip.

Illus. 2-16. The multiple-speed digital readout on the Porter-Cable router gives the speed selected.

Illus. 2-17. The speed-selector dial on the Sears variable-speed plunge router is within your finger's reach from the hand grip.

Illus. 2-18. The speed-control dial on the Elu router has five numbered stages: 1 = 8,000 rpm; 2 = 12,000 rpm; 3 = 16,000 rpm; 4 = 18,000 rpm; and 5 = 20,000 rpm.

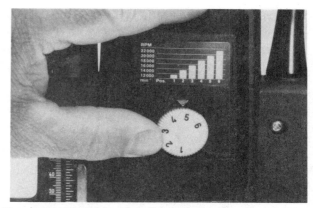

Illus. 2-19. The six-position speed dial on the 1¼ and 2 hp Bosch electronic variable-speed routers. Note how the numbers translate to specific speeds on the chart/scale above.

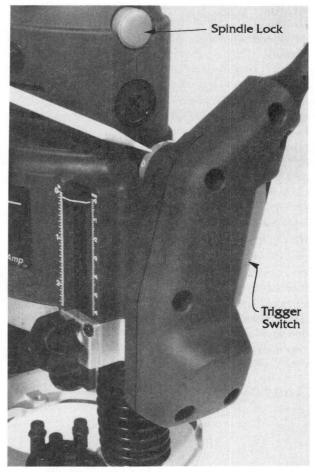

Illus. 2-20. The speed-control dial of the new Sears, 2 hp industrial plunge router is located on the right handle, which also houses the on-off trigger switch.

Sears and RYOBI industrial routers. You can actually change speeds during an operation without moving your hand from the router hand grip (Illus. 2-17 and 2-20).

The new electronic variable-speed circuitry, designated as EVS, offers much more than just convenient speed changing. It features a "soft" or "ramp" start that brings the motor gradually and more safely up to speed rather than in one, big instantaneous starting thrust that can yank the tool from your grip. The new electronics circuitry also maintains the set rpm when the router is under loads that might otherwise make it run at slower speeds. Extra voltage automatically provides the extra power that is needed to maintain the designated speed when the router is making tough cuts.

Bearings, Brushes, and Motor Housings

Most experts agree that the ball bearings on a router are indicative of its structural quality. Cheaper routers generally have cast plastic motor housings and bases (Illus. 2-10). Seating bearings precisely and maintaining them in cast-plastic sockets is by and large not as good as seating them in accurately machined metal housings. Bearings must endure the constant sideways forces or thrust pressures against the motor shaft. The heavier and larger the bearings, the longer the life of the router. In a 1992 article in *Fine Woodworking*, Robert Vaughan compared the bearings on different routers. Various Elu, Porter-Cable, Freud, and Bosch routers were determined to have the largest bearings. Certain RYOBI and Hitachi routers had proportionally smaller bearings.

Metal housings (Illus. 2-12) are preferred over cast-plastic, for a couple of reasons. First, metal-housed motor units are more durable. Second, these motor units can be used in more clamping and holding applications when just a router's motor is being used as a power source in one of the accessories now available. The motor brushes should be easily accessible so that they can be replaced by the user fairly easily.

Cords

Power cords may seem unimportant to router usage, but certain cord arrangements can be very annoying. First, on all non-D-handle-type routers I prefer cords that exit from the motor in a straight-up, vertical direction. Cords that come out horizontally from the motor (like those on Skil, some Black & Decker, Sears, and other routers)

always seem to be in the way. The best power cord arrangement is a cord that exits through a D handle near the bottom. Thus, the cord is more or less always under your arm and not entangled with it.

Cords that are stiff and nonflexible are irritating. Also, some black rubber cords will leave scuff marks on certain woods if slightly rubbed across corners, the edges of boards, and/or furniture and cabinet assemblies. Finally, undersized power cords should have adequate "strain relievers" at areas of normal sharp bends, such as at the motor housing or where tension may cause premature cord wear or damage.

Switches and Handles

Power switches located on the handles or hand grips of the router are the safest and certainly the easiest type of power switch to use. The ideal switch is the trigger-type switch that can be activated only by first depressing an auxiliary "lock-off" button usually located to the side of

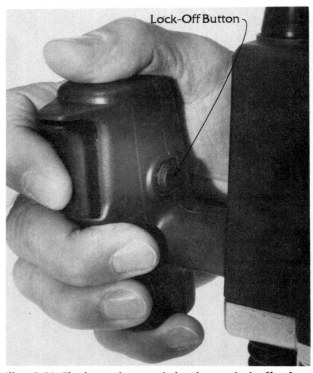

Illus. 2-21. The best trigger switches have a lock-off safety button so the trigger cannot be activated accidentally. The lock-off button must be depressed to permit the trigger to work. Note the contoured design of the handle on this new Bosch plunge router.

the trigger. This auxiliary safety feature reduces the possibility of an inadvertent start-up when you grab the router by the same knob or handle. Bosch and some RYOBI plunge routers have this switching feature (Illus. 2-21).

Toggle, rocker, and slide switches are other types of switches usually located somewhere on the motor unit. They are less preferable than the trigger-type switches. However, if you can reach a switch of this type with your finger without removing your hand, it is probably safe to use (Illus. 2-22–2-24). The worst switches are those that are located so far away you must remove or change your grip to reach them. Some fixed-base routers with poor switch locations are shown in Illus. 2-25 and 2-26.

Though you should determine the type of switch to use according to your general routing needs and expectations, the switch should be easily operable and easily accessible so that it can be quickly shut down should some problem arise during use. If the nature of your work involves a lot of starting and stopping, as would occur in freehand work, carving, and dovetailing, a switch that is easily accessible is of major importance. If you already are using a router with a poorly accessible switch, consider augmenting your system with a foot-operated switch accessory.

Perhaps the best system on fixed-base routers is the kind with two switches, that is, one switch located in the grip or handle of the router base and another on the motor itself (Illus. 2-28). Consequently, when you want to

Illus. 2-23. The vertical slide switch on the Hitachi router is also usually reachable from the handle.

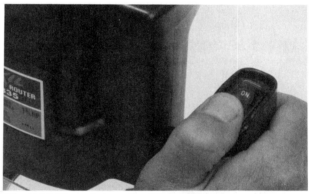

Illus. 2-24. The switch on the Skil small plunge router is located on top of the handle grip, and is covered with a flexible dust cover.

Illus. 2-22. The power switch and speed control on the RYOBI 3 hp plunge router is usually within reach of the operator's finger.

Illus. 2-25. Depending upon the depth adjustment, the toggle switch on Elu, DeWalt, and some Black & Decker fixed-base routers is seldom reachable from the knobs.

Illus. 2-26. The toggle switch on the Porter-Cable 1 and 1½ hp routers cannot always be reached from the knobs or handles.

Illus. 2-28. The author's favorite type of fixed-base router has the following special features: the cord exits near the bottom of a D-style handle; it has two switches, one on the motor unit and a spring-loaded trigger switch in the D-handle; and its handle and knob are positioned low.

Illus. 2-27. Black & Decker's consumer router. Note the motor switch, integral cast plastic base, and handles. The switch can be located various distances from the handle, depending upon the vertical depth-of-cut adjustment of the motor unit.

Illus. 2-29. This type of handle design can be found on Elu, DeWalt, and Black & Decker fixed-base routers.

use the motor independently of the base, the router motor has its own switch. This is especially practical when you are using the router motor unit in a router table, overarm, or in one of the many other accessory devices now available.

Handle Designs

There are many types of handle design. This includes different grip, knob, and handle shapes. The grips on plunge routers are used to push the router down to the desired depth setting, and then to guide the router forward during the cut. Hitachi plunge routers have adjustable handles that can be tilted to accommodate individual preference. Some small routers have slippery knobs (Illus. 2-29). This may be a cause for concern.

Though determining which type of handle to use is a matter of personal preference, it is an important decision, especially when you are selecting a professional router for more than casual or short term use. Select a handle contoured for comfort. Look at the variety of handle designs shown in Illus. 2-29–2-33. Other factors to consider when choosing a handle design are where the handles are placed on the router and how they are attached. Some of the less expensive routers have non-removable handles or knobs integrally moulded into motor housings and/or bases (Illus. 2-30 and 2-33). You may prefer removable knobs and handles. Otherwise, they might be troublesome when you are using the router with

accessories such as router tables or when making other routing devices of your own design.

Should the knobs or handles be located higher up or lower down on the router base? Each location has advantages in certain situations. Personally, I like them low on the router base. In freehand work, this allows me to press my wrist and forearms down on the surface of the work, which gives me better overall control. I can use my wrist as a pivot by making a compass-like maneuver for small curves, or stiffen my wrist and pivot from my elbow when making larger curves.

Handles and knobs located low on the router base can interfere when you are guiding the router along some straight edges and in certain jig and fixture work. Plunge routers feel top-heavy and feel as if they are ready to tip over, especially when they are in the up, retracted position (Illus. 2-1 and 2-34).

For some work such as fast and light sign-making and freehand carving, routers with no knobs or handles may be preferable (Illus. 2-35).

Collets

The design and size of the collet, the part on the motor spindle (shaft) that grips the bit, is an extremely important feature to be concerned about. A good collet system must, over the long term, always hold the bit true and concentric to the center axis of the motor shaft so that it rotates without the slightest wobble or vibration. It must

Illus. 2-30. These routers have handles with built-in trigger switches. On the left is a Black & Decker fixed-base router. On the right is a RYOBI plunge router.

Illus. 2-31. The handles on the Hitachi router can be adjusted to any of three angular positions for best operator comfort.

Illus. 2-33. Note the two radically different designs of the Skil router (left) and the Elu router (right). Note the handles integrally moulded with the all-plastic motor housing, the exposed internal springs, and the opening inside the base.

Illus. 2-34. Most plunge routers feel "top-heavy" when compared to fixed-base routers.

Illus. 2-32. This Elu 1 hp plunge router has an unusual handle that was probably designed to accommodate the use of a special right-angle base accessory used for edge work, with the router held horizontally rather than vertically. Note the switch location.

Illus. 2-35. The small-diameter motors of trim routers are convenient for one-hand operation; thus, no knobs or handles are necessary.

be of a substantial design intended for long-term wear and performance, while maintaining its very stringent dimensional tolerances. Any tolerance deviation in terms of bit wobble (runout) places stress upon the bearings, creates tool vibrations, collet wear, and premature bit wear, and retards the bit's cutting efficiency. This produces poor cuts and can even lead to a hazardous operating situation.

The collet should be long enough to grip a good length of the bit's shank; in fact, the longer the collet the better (Illus. 2-36). Smaller routers have collets that will only accept ¼″ diameter bit shanks. Larger, higher-priced routers have collet capacities of ½″. A router with the capability to use bits with the larger (½″) shank diameter has an advantage, for many reasons. In addition to using bits that are stronger and stiffer, you can also use a broader choice of cutters with the larger collet. This includes all production-type bits, which offer you more selection in cutting configuration possibilities.

Present-day routers have a wide variety of collet lengths and designs (Illus. 2-37 and 2-38). These collets range from those under ½″ in length to those over 1½″. This is a considerable difference in the size of the part of the router that is actually gripping the router bit, which is spinning at a high speed. Collets also have different numbers of slits. There are small collets with just one compressive slit and ½″ collets (Illus. 2-39–2-41) with as many as eight multiple slits, which provide more uniformly concentric compression (Illus. 2-42). Good ¼″ collets have four slots. The shanks of cheaper, imported bits may be slightly undersize, making them difficult to grip adequately with the lower-quality collets.

Many economy (low-to-medium grade) collets have a 3-slit design (Illus. 2-38 and 2-39) and may also not be adequately tempered. A properly tempered collet does not wear or "bell" out, causing concentricity problems. To test a collet to be sure it's properly tempered, just run the edge of a file against it. If it makes a cut, the steel isn't tempered and you'd be wise to look for a better, more serviceable collet. Once the lock nut on a good collet is

released, the collet will spring open, permitting the bit to slip out. Poorer-quality collets sometimes remain pinched against the bit. This makes it almost impossible to remove the bit.

The new, better-quality routers feature "self-releasing collets." With these types of collets, you are spared the frequent frustration of dealing with difficult-to-remove bits. As you loosen the lock nut, a "keeper ring" pulls the collet out with it, thus disconnecting the bit and collet from the router's motor shaft (spindle).

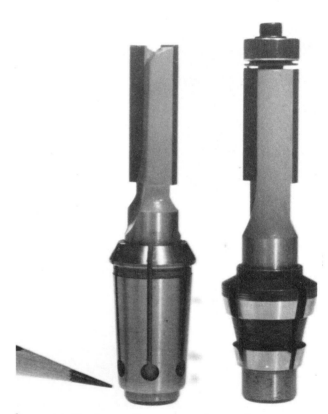

Illus. 2-36. **The best collets are very long, yet compress as uniformly and as concentrically as possible around the shank and spindle of the router.**

Illus. 2-37. **This sample of various collets indicates the wide variety of overall sizes and designs found on present-day routers. Note the shortest collet at the far left, with just one slit.**

Illus. 2-38. The ¼″ collets on the Sears light-duty routers are bored directly into the end of the slotted and threaded motor shaft.

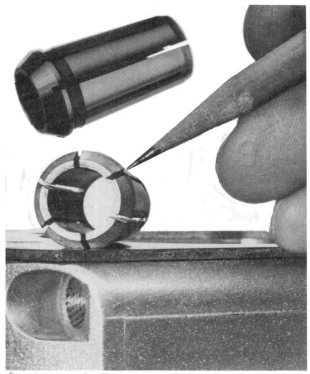

Illus. 2-39. Good collets are not only long, but have more slits for more uniform compression around the shank of the bit. The insert shows DeWalt's collet, which is slotted on 8 sides. Good collets are also made of durable, wear-resistant tempered steel.

Illus. 2-40. Left: How a 3-slit collet grabs the bit. Notice that there are fewer actual contact areas to compress against the bit than the more uniformly concentric compression provided by collets with more slits, as shown on the right.

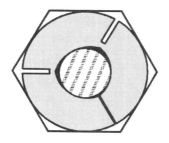
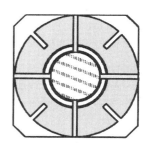

Illus. 2-41. A close-up look at three good collets and lock nuts found on heavy-duty plunge routers. From left to right, these collets and lock nuts are found on Elu, Bosch, and Makita routers.

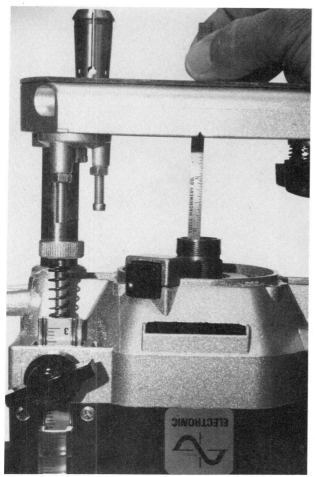

Illus. 2-42. Elu's big plunge routers have the longest collets (1¾″) and deepest openings of any router available today.

Illus. 2-43. Freud's router features a spring-loaded sliding shaft lock.

Illus. 2-44. A push-button shaft lock on the Elu plunge router.

Shaft Locks

Shaft Locks (Illus. 2-43 and 2-44) are a feature that appears to be gaining popularity. Typically, many routers have come with two wrenches, one to grab the motor shaft and the other to grip the collet lock nut. Most manufacturers are now switching to the locking-shaft, one-wrench system. However, some professionals still insist the old way is better. Their reasoning relates to their ability to manipulate two wrenches with just one hand to tighten or loosen the collet.

Spindle shaft, or arbor locks as they are called, have different activating mechanisms. Most feature a spring-type mechanism that is activated by a push button, slide, or lever pivot device that holds the motor shaft rigid as the single wrench is applied to the collet lock nut.

Bases and Sub-Bases

The sub-base of a router, which is the bottom of the base, can be round, square, or round with one or two flats (Illus. 2-45). Each shape has certain advantages and limitations, but the differences between shapes are basically minor. Bigger and round sub-bases seem to provide more support for edge-routing jobs. Routers with sub-bases having a flat or straight edge are supposedly best for following straight-edged guides and allow cuts to be made closer to vertical obstructions.

There are other features of sub-base design that are more important. First, the plastic sub-bases should be removable so that they can be replaced with special shop-

Illus. 2-45. Left: The standard sub-bases on the Porter-Cable fixed-base and plunge routers have basically the same design. Right: The sub-base on the small Elu is not removable.

made sub-bases designed for specific routing jobs. It is very puzzling to me why standard, factory sub-bases are not made from a durable, clear plastic. The typical black sub-bases cannot be looked through. For years I've made my own (Illus. 2-46). At least one manufacturer, Porter-Cable, now provides clear bases as accessories, but they are not a substitute for the company's regular bases, which are made to carry template guide inserts (Illus. 2-47). Template guides are tubular attachments mounted to sub-bases that allow the router to follow specific templates, patterns, or special jigs. They are discussed in Chapter 13.

The round sub-bases on some routers are more precisely manufactured than are others. The circumference edge of the most precisely manufactured round sub-bases is concentric to the central axis of the bit. Thus, every point along the circumference of this sub-base is exactly the same distance from the bit.

Illus. 2-47. Many foreign and United States manufacturers provide special base inserts that are made to receive Porter-Cable's template guides (below), considered by many to be the standard in the industry.

Illus. 2-46. A view of one of my shop-made, clear-plastic sub-bases. Notice the large center hole and the slightly rounded edges on the plastic base. Use polycarbonate-type plastic. It's tough and virtually scratch- and mar-proof. Avoid using acrylic plastics. They are more difficult to machine and not nearly as tough.

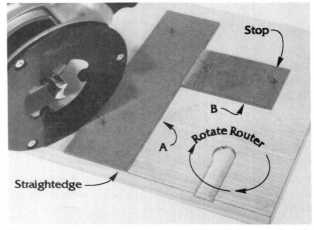

Illus. 2-48. Testing the concentricity of the sub-base to the bit. While keeping the router base against the straightedge (A) and stop (B), rotate the router 360 degrees. The size of the cut will equal the cutting diameter of the bit if the base is concentric. If not, the cut will be larger than the bit.

An easy test for concentricity of any round router sub-base is shown in Illus. 2-48. Feed the router a short distance along a straightedge to a stop. While keeping the router against the straightedge and against the stop, rotate it 360 degrees. Then measure your cut and compare it to the size of the bit used. If they are equal, your round base is concentric to the bit.

Another very important consideration is how easily, if at all, the router's base can be secured to a shop-made or commercially produced router table or similar base-mounted accessory device. Are sufficient screw or threaded sockets provided in the base (Illus. 2-49)? Some of the best, most expensive routers lack this basic feature.

Illus. 2-50. A typical adjustable edge-guide accessory that fits to the router base.

Illus. 2-49. A bottom view of a Sears industrial plunge router with its sub-base removed. Note the four large threaded sockets which are sufficiently sturdy for inverted mounting in a router table.

The base is also where many other accessories are attached to the router. Various edge guides (Illus. 2-50), fences, circle-cutting devices, template guide bushings, trimming guides, lights (Illus. 2-51), dust-extraction and other attachments are held to router bases by slotted screws or thumbscrews. All devices should interchange easily, and tighten securely. Determine if the mounting areas are sufficient and if stripped threads are likely to occur because the housings are made of soft metal or plastic. One very helpful device is a clear-plastic chip shield that is mounted to the operator's side of the base.

Serious consideration should also be given to the size of the opening in the base. It's nice on flat surface work to have lots of space around the bit that is visible to the operator. This is especially important when following irregular layout lines freehand. However, large openings may provide serious support problems when you are

Illus. 2-51. Some routers feature a light inside the base, which is helpful in freehand router work for following layout lines.

mortising and routing edges of narrow stock by hand or shaping edges at the corners of boards. The size of the hole diameter through the cast base and the attached sub-base determines the maximum cutting diameter or size bit you can use in the router. Although the trend is towards using bits of dangerously large diameters, many routers do not allow the use of moderately sized bits even in the 1¼″ diameter range. Conversely, other routers have very large openings that, with the factory sub-base still attached, can accommodate bits with 3″ diameters. In my opinion, the hazards of routing increase proportionately with the diameter of the bits used, and warnings concerning this will be made frequently in future chapters.

Dust Collection

Removing the shavings and collecting the microscopic fine dust particles from the routing area and shop environment still remains one of the most serious problems associated with routing. Efficient dust collection, especially for all of the various hand-held jobs, is still not accounted for in present-day router designs. Some router manufacturers do offer dust-collecting accessories; most are hoses that attach to shop vacuums (Illus. 2-52–2-57). The dust-collection system on the Elu router is the best-designed commercial system I've seen. Because it is vertical and fits alongside the router, it does not interfere with normal routing operation, unlike other dust-collection systems. (Refer to Chapter 8 for more information on dust collection).

Noise Levels

The new electronic routers seem to be somewhat quieter than their earlier counterparts. Still, the noise level of all routers is far too high. According to an article in *Furniture Design and Manufacturing* magazine, AEG and Elu have the quietest routers, and Hitachi, Makita, and RYOBI the noisiest. All routers today still do pose a very serious risk to hearing if used without hearing protection. In this regard, it is my contention that routers should be required to carry warning labels along with sound-level ratings. (Refer to Chapter 8 for information about hearing protection.)

Illus. 2-53. The chip collector on the Milwaukee router is similar in design and function to those of several other fixed-base routers, including the Bosch and Porter-Cable routers.

Illus. 2-54. The vacuum connection for dust pick-up on the Hitachi router projects at a low angle.

Illus. 2-52. The Sears Craftsman 2 hp electronic router features a dust pick-up system that draws up one handle into a bag.

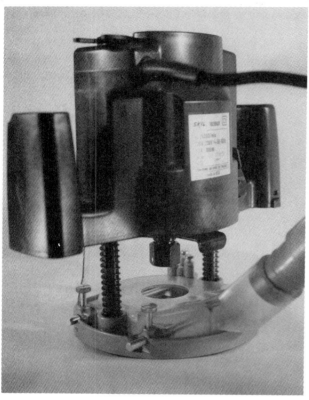

Illus. 2-55. Skil's dust-collecting attachment is similar to some other attachments.

Illus. 2-57. Elu's vacuum pick-up runs up vertically and fits neatly alongside the router.

Illus. 2-56. AEG's dust pick-up accessory.

Net Weight

The physical weight of routers varies from approximately 4 pounds to over 17 pounds. Heavier routers are generally more stable and more shock-absorbant than lighter ones that do the same jobs. They are good for router table setups and similar long-term setups. Jobs requiring much lifting, reaching, and hand-held portable routing work are obviously less tiring when done with lighter routers.

Air-Powered Routers

Air-powered routers (Illus. 2-58 and 2-59) are certainly well worth considering, if you have a suitable air supply available. Air or pneumatic routers have some advantages over their electric counterparts. They are generally similar in appearance, do most of the same jobs as electric fixed-base routers, and are available with some of the basic attachments such as template guides and laminate trimmers. Their advantages are as follow:

1. They are generally considerably lighter in overall weight than electric routers of equal horsepower and speed capacities.
2. They can reach higher rpm—up to 40,000 rpm with ¼″ collets and as high as 65,000 rpm with smaller, ⅛″ collets.
3. They operate at lower temperatures and, therefore, are more comfortable to use continuously.
4. They are free of any electrical parts; consequently, they may be used safely in wet or damp environments.
5. They have fewer working parts and are less expensive to maintain and repair.
6. Usually, no lubrication is required.
7. Their motors cannot burn out. Even when overloaded to the point of full stall, air motors are not damaged.
8. They are generally quieter in operation.

One disadvantage of using air-powered routers is that they require a substantial-sized air compressor. It must provide a continuous flow of air with an output range of 10–38 cubic feet (0.5–1.0 cubic metres) per minute (depending upon router horsepower size) at 90 pounds per square inch. They also must be used with solid-carbide bits, rather than brazed bits which could fly apart at their high speeds. Additionally, the following are required: in-line moisture filters to dry the air supply; a lubricator; and a pressure regulator to ensure that the air supply is properly conditioned and will not cause internal damage to the tool.

Buying a Router

Before buying a router, be sure to review the features that are common to both fixed-base and plunge routers, as discussed in this chapter. Chapters 3 and 4 also discuss features that apply to the specific types of router, such as adjustments, stops, and clamping systems. They also provide tables that compare the features of the same type of router.

Do the following before buying a router: (1) obtain opinions from friends or colleagues who own the model routers you are considering; (2) read current periodicals and their tool reviews; and (3) test the tool yourself by performing your own kind of work. Simply relying on another's opinion is not enough. That person may have entirely different uses and expectations for the router than you do.

Illus. 2-58. A typical air-powered router fitted with a template guide.

Illus. 2-59. This lightweight, air-powered router by Air Turbine Technology, Inc., features a trigger throttle speed control and a ⅝ hp, ¼″ collet, 40,000 rpm motor that is quiet, running at less than 78 decibels.

3
Fixed-Base Routers

This chapter examines and compares some features exclusive to the fixed-base routers. Included are illustrations from the major manufacturers of fixed-base routers and a table that compares the features of current routers (Table 3-1 on pages 40–42).

As mentioned earlier, the motor units on most fixed-base routers can be removed from their bases (Illus. 3-1 and 3-2). However, this is not always the case. The Sears Craftsman consumer routers are "wired" together from the motor to their handle trigger switch, as shown in Illus. 3-3. This makes it very difficult to remove and use the router motor unit independently with self-made fixtures and other non-Sears accessories. Most of the Sears accessories are designed to accept the entire router, so for these accessories, at least, the motors do not have to be removed from their bases.

Illus. 3-2. The Milwaukee 1, 1½, and 2 hp fixed-based routers have the same design and overall appearance.

Illus. 3-1. The motor units on most medium- and heavy-duty fixed-base routers can be removed from their base. Note the switch and flat top design characteristic of all Milwaukee routers.

Illus. 3-3. Typical of all Sears Craftsman consumer routers is this non-removable power-cord connection from the motor to the handle switch.

FIXED-BASE ROUTERS

MFG	Model No.	Maximum Collet Capacity (inches)	Amps hp Rating	Non-Load rpm	Motor Shaft Lock	Switch T = Trigger O = Other	Weight (lbs.)
Black & Decker	7604	¼	5 amp 1 hp	30,000	No	O	4.1
	7613	¼	8.5 amp 1¼ hp	25,000	Yes	T	8.1
	7612	¼	9 amp 1½ hp	25,000	Yes	T	10
	TS100	¼	9 amp 1½ hp	25,000	Yes	T	8.1
	2720	¼	8 amp 1½ hp	25,000	No	O	7¼
Bosch	1601	¼	7 amp 1 hp	25,500	No	O	6¼
	1602	⅜	9 amp 1½ hp	25,000	No	O	7
	1604	½	10 amp 1¾ hp	25,000	No	O	7.75
	1606 D Handle	½	10 amp 1¾ hp	25,000	No	T	7.8
	1600 D Handle	½	12 amp 2¼ hp	26,000	No	T	12¾
	90300	½	15 amp 3¼ hp	22,000	No	O	15½
Craftsman (Sears)	17471	¼	1½ hp		Yes	T	7.56
	17472	¼	1¾ hp	15,000-25,000	Yes	T	8.3
	17473	¼	2 hp	15,000-25,000	Yes	T	9.13
	17470	¼	7 amp 1⅛ hp	25,000	Yes	O	4½
Chicago (Harbor Freight)	344 D Handle	½	2 hp	23,000	No	T	8
DeWalt	610	½	9 amp 1½ hp	25,000	No	O	7.3

Table 3-1. Comparison of fixed-base routers. The table continues on pages 41 and 42.

FIXED-BASE ROUTERS

MFG	Model No.	Maximum Collet Capacity (inches)	Amps hp Rating	Non-Load rpm	Motor Shaft Lock	Switch T = Trigger O = Other	Weight (lbs.)
Dremel Moto-Tools	3952 3950 2850	⅛	1.15 amp	Variable 5,000- 30,000	Yes	O	—
Elu	2721	½	9 amp 1½ hp	25,000	No	O	7¼
	3328	½	13 amp 3½ hp	24,000	No	O	14½
Jepson (made) in Taiwan)	7210	⅜	¾ hp	23,000	No	T	5
	7112	½	2 hp	25,500	No	T	10.3
Makita	3606	¼	7 amp 1 hp	30,000	Yes	O	5.5
	3601B	½	8.5 amp	23,000	No	T	8
Milwaukee	5620	⅜	8 amp 1 hp	23,000	No	O	8
	5660	½	10 amp 1½ hp	24,500	No	O	8½
	5680	½	12 amp 2 hp	26,000	No	O	8¾
Porter- Cable	100	¼	6.5 amp ⅞ hp	22,000	No	O	6¾
	630	½	6.8 amp 1 hp	22,000	No	O	7½
	690	½	10 amp 1½ hp	23,000	No	O	8
	691 D Handle	½	10 amp 1½ hp	23,000	No	T	9¼
	7537 D Handle	½	13 amp 2½ hp	21,000	No	T	14½
	7536	½	13 amp 2½ hp	21,000	No	O	14½
	7519	½	15 amp 3¼ hp	21,000	No	O	15
	7518	½	15 amp 3¼ hp	10,000 13,000 16,000 19,000 21,000	No	O	15

Table 3-1 continued. See the following page.

FIXED-BASE ROUTERS

MFG	Model No.	Maximum Collet Capacity (inches)	Amps hp Rating	Non-Load rpm	Motor Shaft Lock	Switch T = Trigger O = Other	Weight (lbs.)
RYOBI	R30	¼	3.8 amp ¾ hp	29,000	No	O	5.6
	R230	½	9.5 amp 1.5 hp	24,000	No	O	12.1
	R331 D Handle	½	12.1 amp 2 hp	24,000	No	T	12.1

Table 3-1.

Depth Adjustment and Base Clamps

All fixed-base routers have mechanisms to clamp the motor in the router's base at a designated vertical position. To effect the desired depth of cut (the amount of bit projecting through the sub-base), you have to move the router motor up or down and clamp it (Illus. 3-4). There are three basic ways motors can be adjusted in router bases: (1) an adjustable ring-and-slide system (Illus. 3-5 and 3-6); (2) a screw-in or spiral system (Illus. 3-7 and 3-8); and (3) a rack-and-pinion system (Illus. 3-9). Most manufacturers favor one system over the others, but you may find manufacturers that use more than one system.

Most of the older Sears and the new EZ routers (Illus. 3-10) use essentially the same clamping mechanism. The parts of the adjustable ring-and-slide system on the

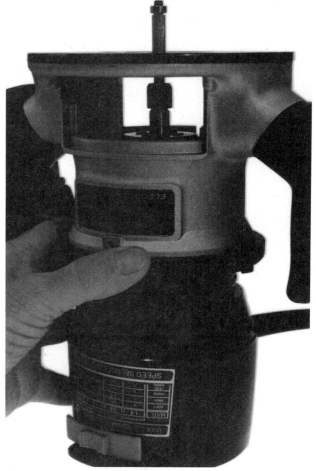

Illus. 3-5. On older Sears routers, a threaded stop ring on the motor unit permits the base to be raised or lowered to achieve the desired amount of bit protrusion through the sub-base.

Illus. 3-4. The base clamp on older Sears models holds the motor at the correct height.

Illus. 3-6 (above). On the new Sears Craftsman EZ routers, the depth lock clamp knob is on an inclined ring which lifts or lowers the motor unit in the base. Illus. 3-7 (right). Twist or screw-in motor systems such as the one on this Porter-Cable router also facilitate the making of micro depth-of-cut adjustments.

Illus. 3-8. Other routers with the screw-in or twist depth-of-cut adjustments. Shown from left to right are Porter-Cable, Bosch, and Milwaukee routers.

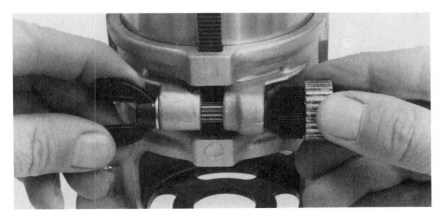

Illus. 3-9. Black & Decker's time-tested rack-and-pinion mechanism makes precise vertical adjustments quickly and easily.

Illus. 3-10. The new light-duty routers from Sears Craftsman are called EZ routers. These routers have a new depth-adjustment system and a spindle lock with built-in storage space for the wrenches.

routers are basically integrated into one functional unit. This system on the earlier Sears routers comprises separate parts. The four new EZ routers all have ¼" collet capacities and range in horsepower from 1⅛ to 2.

Porter-Cable, Bosch, and Milwaukee make considerable use of the screw or twist-in systems. Makita, Jepson, and RYOBI have just one or two fixed-base routers; they utilize the screw-in system as well. Illus. 3-11 shows the

popular model 100 Porter-Cable router, one of the company's oldest, most popular models. Illus. 3-12–3-15 show other Porter-Cable routers.

Illus. 3-12. This Porter-Cable model 690, 1½ hp router looks similar to the model 630 router, except that this one has a ½" collet capacity.

Illus. 3-11. Porter-Cable's model 100, ⅞ hp, ¼" router.

Illus. 3-13. This Porter-Cable model 7519, 3½ hp production router is similar in appearance to the model 7536, 2½ hp router.

Illus. 3-14. This Porter-Cable model 691, 1½ hp, ½" collet router has features this author finds very appealing. They include a D-handle, a trigger switch, and additional removable knobs.

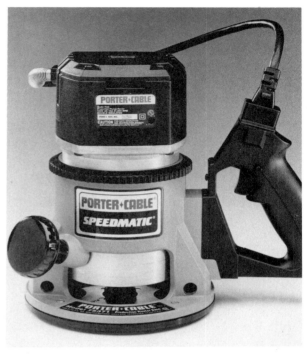

Illus. 3-15. This Porter-Cable model 7537, 2½ hp production router has a D handle.

Overview of Fixed-Base Routers

Bosch makes six fixed-base routers. Three models—1601, 1602, and 1604 (1, 1½, and 1¾ hp, respectively)—have similar designs. The 1604 has a ½" collet, and the 1601 and 1602 models have a ¼" collet. Bosch currently manufactures two routers with D handles: the model 1606 1¾ hp router shown in Illus. 3-16, and the model 1600 2¼ hp router shown in Illus. 3-18. The largest fixed-base Bosch router is the model 90300 3¼ hp router (Illus. 3-19).

The light-duty consumer router line produced by

Illus. 3-16. A variety of routers with D-handles. Top row: Porter-Cable and RYOBI routers. Bottom row, from left to right: Jepson, Makita, and Bosch routers.

Illus. 3-17. The Bosch model 1604 router.

Illus. 3-19. The Bosch model 90300, 3¼ hp router.

Illus. 3-18. The Bosch model 1600, general-industrial 2¼ hp router has a D handle.

Black & Decker has undergone some changes in recent years. Dropped from production are the company's model 7600, 7605, 7614, and 7666 routers. A model TS100 router replaces the model 7614 router. Black & Decker still has a large selection of 1–1½ hp routers to choose from (Illus. 3-20 and 3-21).

Recently, the industrial division of Black & Decker joined forces with Elu and released a new line under the DeWalt name. This line includes some former Black & Decker industrial models that have been improved and now carry either Elu or DeWalt labels. There are slight mechanical changes, but virtually no visible design changes, between the new and old routers. The current line-up of Black & Decker, Elu, and DeWalt fixed-base routers for the industrial-construction market have some very similar features (Illus. 3-22).

The RYOBI and Makita fixed-base routers are shown in Illus. 3-16. Some laminate trimmers are shown in Illus. 3-24, for your information. Bosch currently has five different models, and all have essentially the same motor unit (5.6 amps, 30,000 rpm), but have different bases designed to do a variety of special jobs associated with plastic laminate work. Likewise, Porter-Cable offers three different motor styles with a number of specialty bases designed to do almost any conceivable trimming job. RYOBI has one small ¾ hp motor, which runs at 29,000 rpm and has a ¼″ collet. This motor fits two

laminate trimmer bases and also fits a solid die-cast aluminum base with two side knobs to function just like a regular fixed-base router. Incidentally, Bosch, RYOBI, and Porter-Cable all offer tilting bases to fit their small laminate trimmer routers.

Illus. 3-20. The Black and Decker model 7604, 1 hp, ¼″ collet router with a ring-type depth adjustment. Note the switch on the motor.

Illus. 3-21. Black and Decker's consumer router model 7613 has a ¼″ collet, a spindle lock, and a trigger switch. The model 7612 router has the same features.

Illus. 3-22. Features of Black & Decker's model 2720, 1½ hp, ¼″ collet router, shown at left, have been incorporated into the new, more powerful Elu model 2721 router (B) and DeWalt's model 610 router (C). Both routers feature Elu's high-quality ½″ collet system.

Illus. 3-23. Elu's model 3328 router is the company's most powerful (3½ hp) router. It is based on a former Black & Decker router and has a ½" long collet with 6 slits and a double D handle design. Here it is cutting through a ¾" particleboard sink opening in just one pass.

Illus. 3-24. Laminate trimmers are small, high-speed routers with a ¼" collet but without handles. They fall into the fixed-base category. Shown here, from left to right, are Bosch, RYOBI, and two Porter-Cable laminate trimmers.

Illus. 3-25. RYOBI's small laminate trimmer motor is the same power source used in the company's small, model R-30 fixed-base router.

4
Plunge Routers

In the late 1970s, plunge routers still were unheard of in most of the United States. When the first edition of this book was published in 1983, imported plunge routers were just starting to appear in United States magazines and mail order catalogues. Today, there are more plunge routers available than fixed-base routers. The high-quality plunge routers are loaded with features that make them more "user friendly" than ever (Illus. 4-1).

Plunge routers continue to be popular because the plunging function gives router users one important advantage over those who use fixed-based routers: the base is spring-loaded to the motor system; this allows the motor unit to be pressed down so that the bit can be lowered to the stock so as to enter the work at exactly 90

degrees. With just a flip of a lever (or similar mechanism), the cutter is automatically retracted at the completion of the cut. Unlike cuts made with fixed-base, hand-held routers, plunge cuts can be started and completed without lifting the base of the router off the work surface. This offers obvious safety features, because the bit protrudes below the base only during the actual routing operation.

Changes in cutting depths can be made quickly, even with the power still on, if desired. This is especially handy when making cuts that are too deep to be made in a single pass.

With the motor clamped so the bit is at the desired depth, the plunge router is used just like a fixed-base

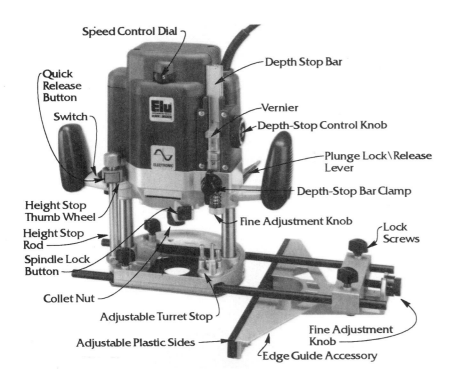

Illus. 4-1. Elu has set the standard for plunge routers for many years. Shown here is Elu's best router: an electronic 2¼ hp, 8,000 to 20,000 rpm variable-speed router.

PLUNGE ROUTERS

MFG	Model No.	Maximum Collet Capacity (inches)	Amps hp Rating	Non-Load rpm	Motor Shaft Lock	Switch T = Trigger O = Other	Max. Plunge Depth (inches)	Weight (lbs.)
*AEG	OFS 50	¼	6.25 amp 1 hp	25,000	Yes	O	1¹⁵⁄₁₆	5.5
	FSE 850							
Black & Decker	7615	¼	9 amp 1½ hp	25,000	Yes	T	2	8¼
	6200	¼	5.4 amp 1¼ hp	24,000	No	O	1¹⁵⁄₁₆	6.5
Bosch	1614	¼	7.5 amp 1 hp	27,000	Yes	T	2	7.8
	1614 (EVS)	¼	7.8 amp 1¼ hp	12,000-23,000	Yes	T	2	7.9
	1613	½	9.8 amp 1¾ hp	25,000	Yes	T	2	9.8
	1613 (EVS)	½	10.1 amp 2 hp	12,000-23,000	Yes	T	2	10.1
	1615	½	14 amp 3 hp	25,000	Yes	T	3	12
	1615 (EVS)	½	15 amp 3¼ hp	12,000-19,000	Yes	T	3	12¼
Craftsman (Sears)	27505	½	12 amp 2 hp	25,000	Yes	T	2½	11.5
	27506	½	15 amp 3½ hp	10,000-25,000	Yes	T	2⅜	16
DeWalt	614	¼	6.6 amp 1¼ hp	24,000	Yes	O	2½	6
	615	¼	8 amp 1¼ hp	8,000 24,000	Yes	O	2½	6.2
	624	½	15 amp 3 hp	22,000	Yes	O	2¹⁄₁₆	11¼
	625	½	15 amp	8,000 22,000	Yes	O	2⁷⁄₁₆	11¼
	3404 (EVS)	¼	6.5 amp 1¼ hp	8,000-24,000	No	O	1¹⁵⁄₁₆	6.2

*As of late 1993 AEG routers will be discontinued in American markets.
Table 4-1. Comparison of plunge routers. See pages 51 and 52.

PLUNGE ROUTERS

MFG	Model No.	Maximum Collet Capacity (inches)	Amps hp Rating	Non-Load rpm	Motor Shaft Lock	Switch T = Trigger O = Other	Max. Plunge Depth (inches)	Weight (lbs.)
Elu	3303	¼	6.5 amp 1¼ hp	24,000	No	O	1¹⁵⁄₁₆	6
	3404 (EVS)	¼	6.5 amp 1¼ hp	8,000-24,000	No	O	1¹⁵⁄₁₆	6.2
	3337	½	12 amp 2¼ hp	20,000	Yes	O	2⁷⁄₁₆	11¼
	3339 (EVS)	½	15 amp 3 hp	8,000-22,000	Yes	O	2⁷⁄₁₆	11¼
	3338-2 (EVS)	½	8.5/7.0 amp 220V 3 hp	8,000-20,000	Yes	O	2⁷⁄₁₆	11¼
Freud	2000 (EVS)	½	15 amp	22,000	Yes	O	2¾	12.5
Hitachi	M8V (EVS)	¼	7.3 amp 1.5 hp	10,000-25,000	Yes	O	2	6.6
	TR12	½	12.2 amp 3 hp	22,000	No	O	2⅜	11
	M12SA	½	14.6 amp 3 hp	22,000	Yes	O	2⁷⁄₁₆	11½
	M12V (EVS)	½	15 amp 3¼ hp	8,000-20,000	Yes	O	2⁷⁄₁₆	11.7
Jepson (made in Taiwan)	7412	½	3 hp	23,000	Yes	O	2½	13
Makita	3620	¼	7.8 amp 1¼ hp	24,000	No	O	1⅜	5.7
	3612B 3612BR	½	14 amp 3 hp	23,000	Yes	O	2½	12.7
Metabo	PT 508	¼	5 amp ¾ hp	27,000	No	O	2	7¾
Porter-Cable	693	½	10 amp 1½ hp	23,000	No	O	2½	11½
	7538	½	15 amp 3¼ hp	21,000	No	T	3	17¼
	7539	½	15 amp 3¼ hp	10,000 13,000 16,000 19,000 21,000	No	T	3	17¼

Table 4-1 continued. See the following page.

PLUNGE ROUTERS

MFG	Model No.	Maximum Collet Capacity (inches)	Amps hp Rating	Non-Load rpm	Motor Shaft Lock	Switch T = Trigger O = Other	Max. Plunge Depth (inches)	Weight (lbs.)
RYOBI	R50	¼	⅜ amp ¾ hp	29,000	No	O	2¼	5.1
	R150	¼	6.5 amp 1 hp	24,000	Yes	O	2	6.2
	R151	¼	6.5 amp 1 hp	24,000	Yes	T	2	6.2
	R500	½	13.3 amp 2¼ hp	22,000	Yes	O	2⅜	11
	R501	½	13.3 amp 2¼ hp	22,000	Yes	T	2⅜	11
	R600	½	15 amp 3 hp	22,000	Yes	O	2⅜	13.6
	RE600 (EVS)	½	15 amp 3 hp	10,000- 22,000	Yes	O	2⅜	13.6
Skil	1823	¼	8.5 amp 1½ hp	25,000	Yes	O	2	6.9
	**1835	¼	9 amp 1.75 hp	25,000	Yes	O	2	7
	1870	½	12 amp 2.25 hp	23,000	Yes	T	2½	9½
	1875	½	11.5 amp 2¼ hp	10,000 23,000	Yes	T	2½	10.25

**Skil's Model 1835 plunge router is also sold under the Master Mechanic label.
Table 4-1 continued.

router. However, the motor units of plunge routers, as a rule, do not separate from their bases easily, if at all. Thus, motor units from plunge routers cannot be as readily used as the power unit in router accessories such as carvers, some router tables, and similar devices. Fixed-base routers are probably as a rule easier to incorporate into most router tables, but many plunge routers can also be ideal for such use. Before buying a plunge router for portable hand-held use or inverted stationary use, consider the features described below and review Chapter 2.

There are some standard components on plunge routers that fixed-base routers do not have. These include plunge posts and springs, the plunge lock/release mechanisms, depth-setting and fine-adjustment devices, and various up-and-down-stroke stops. The following sections examine these parts and look at repre-sentative plunge routers from current manufacturers. Also, Table 4-1 compares the features of different plunge routers.

Plunge Posts and Springs

The coil springs that lift the motor unit generally fit inside hollow, polished steel posts. The coil springs on some routers fit *outside* the steel posts. These include the Metabo router (Illus. 4-2), the small RYOBI router (Illus. 4-3), the small Skil model 1835 router (Illus. 4-4), and the Master Mechanic router with the same Skil design. The exposed springs make the posts difficult to clean, but otherwise they function much like the internal springs. The new Bosch and Sears plunge routers have flexible, bellows-type covers that protect the posts from sap, pitch and dirt (Illus. 4-5 and 4-6).

Illus. 4-2. The German-made Metabo router has a ¼" collet, a 5-amp, 27,000 rpm motor, and a plunge depth of 2". The motor unit can be removed and used as a die grinder and to power a flexible shaft accessory.

Illus. 4-3. The small, ¾ hp motor from the RYOBI laminate trimmer is the power source for RYOBI's smallest plunge router, the model R-50 router.

Illus. 4-5. The new Sears industrial 3½ hp, 25,000 rpm plunge router features a trigger switch, a shaft lock, and a plunge lock and release that are built into the handle.

Illus. 4-6. The new Bosch 1 and 2 hp electronic variable-speed routers have some distinct design innovations, such as flexible bellows-type post covers, pivoting chip deflectors, and other favorable features.

Illus. 4-4 (left). One of the least expensive routers is this Skil model 1835, 1¾ hp plunge router, which may also be found with a Master Mechanic label in some hardware stores. Note that the plunge return springs are exposed outside the posts.

It is essential that the plunging action of the router you select be smooth, accurate, and consistent. Make sure the plunging action is not sloppy and does not stick. Test it by plunging downwards using just one handle. It should not bind or have any sideways play in its overall plunging performance. Check this carefully, since any two routers of the same brand and model may perform quite differently.

Plunge Lock and Release Systems

Plunge routers have four basic types of mechanism that lock and release the plunging depth. They are: (1) the simple lever action (Illus. 4-7 and 4-8); (2) the separate lock and release actuators, such as those found on the

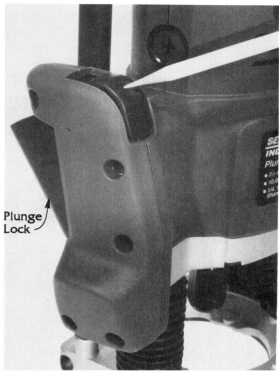

Illus. 4-9. The new Sears industrial plunge router has individual release and lock mechanisms built into the left handle. Depressing the top button releases the plunging depth, and squeezing the trigger-like lever on the handle locks it.

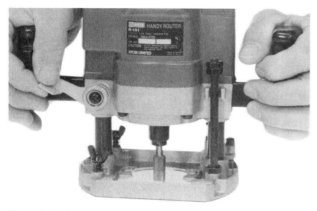

Illus. 4-7. The plunge lock/release lever on some routers is made of stamped metal and has sharp edges. It is too short for good leverage. To use the lever, you must press it downwards to lock it.

Illus. 4-8. The plunge lock/release lever on the big Elu router is a smooth, contoured casting, but it is also too short.

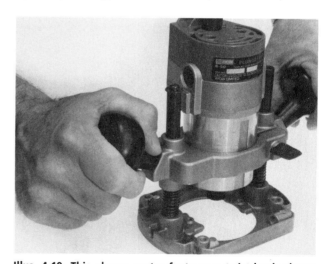

Illus. 4-10. This plunge router features a twist-knob plunge lock. Simply rotating the right-hand knob clamps the motor in place. This type of lock takes a while to get used to. You might find yourself inadvertently loosening the clamp when lifting the router or using it.

new Sears industrial router (Illus. 4-9); (3) the twist knob (Illus. 4-10), as found on the small RYOBI, Elu and DeWalt routers; and (4) the spring-action, self-locking levers (Illus. 4-11 and 4-12), such as those found on the Porter-Cable and Bosch routers. I really enjoy using the

Illus. 4-11. The Porter-Cable router has a spring-loaded, self-locking mechanism that clamps firmly upon release.

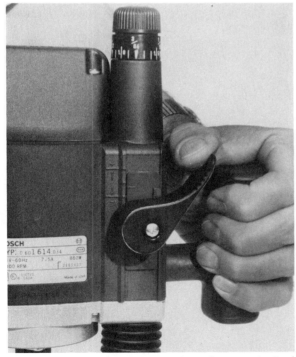

Illus. 4-12. The spring-loaded lock/release lever on the Bosch router is easily operated with the thumb, as shown. It also clamps automatically upon release.

spring-loaded, automatic self-locking levers that are found on the Porter-Cable and Bosch routers. They are definitely the easiest to use. It's too bad manufacturers of otherwise very fine routers continue to provide small levers with sharp edges that are tough on one's fingers after short periods of repeated use.

Depth-Stop Systems

Almost all plunge routers have some type of scale and clampable rod or stop-bar that is located alongside the motor (Illus. 4-13). This is used in conjunction with an incrementally stepped, rotating turret at the base (Illus. 4-13 and 4-14). Some of the devices on the deluxe routers are sophisticated and offer very precise depth-of-cut settings. Most allow you to "zero out" the router, that is, to adjust the scale or vernier at the zero position when the bit is lowered to the point where it just touches the surface of the workpiece (Illus. 4-15–4-17).

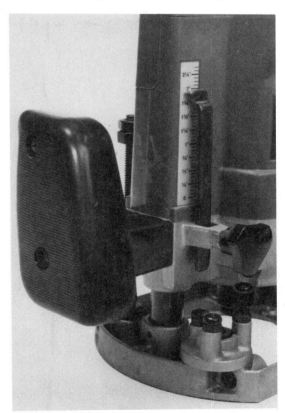

Illus. 4-13. The most basic depth-adjustment-stop system consists simply of a scale and an adjustable stop bar with a three-position turret, such as the one on this small, 1 hp RYOBI plunge router.

Illus. 4-14. This Bosch router has an eight-level turret for presetting multiple depths. Note the tip-out chip deflector to the right and the unusual snap-in design of Bosch's template guide, shown in the foreground.

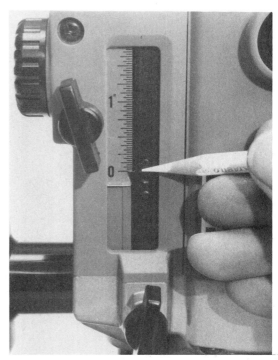

Illus. 4-16. "Zeroing out" the depth of cut. The scale is actually adjusted to read zero when the bit just touches the surface of the workpiece.

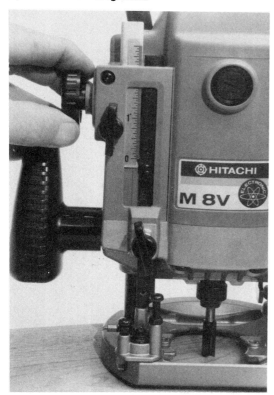

Illus. 4-15. Some routers enable you to "zero out" the depth scale with the bit just touching the work surface. Further adjustments can provide the precise depth of cut that's readable on the scale, which you can divide into multiple pass cuts with the aid of the turret system.

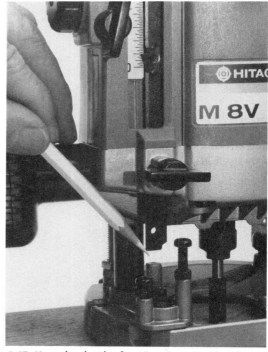

Illus. 4-17. Here the depth of cut is set precisely at ¼".

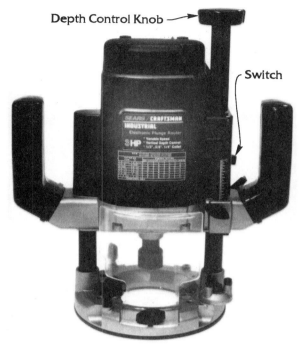

Illus. 4-18. This Sears 3 hp plunge router, made by RYOBI, comes with a depth-control knob which is especially convenient for stationary router-table use.

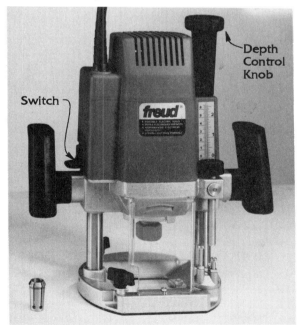

Illus. 4-19. Freud routers come with a depth-control knob that permits a fairly long range of bit movement and can also be used to make micro-adjustments.

Some of the depth-setting devices seem to be over-designed. Unless one requires a very precise inlay or groove when using a portable hand-held router, I question their value. When I cut deep mortises, for example, I rely more on the sound of the motor and often change depths during a cut. These devices will probably prove advantageous for production jobs, but I still prefer to cut to the final depth by setting the stop rod (bar) for the maximum depth I need and then working to the final depth in as few passes as possible without even taking the time to rotate the turret.

A router on which you can make fine adjustments after you've locked everything in may prove beneficial in precise joint work and when the router is used in a router table. The Bosch routers (Illus. 4-6) feature what is referred to as a "microfine bit depth adjustment," which can be used to fine-tune the depth adjustment to within .004″.

One disadvantage of some plunge routers is that they are not ideally suited to be inverted in router tables. The router you have may be difficult to use if it either has a switch in the handle such as a trigger switch that does *not* have a lock-on button or does *not* have depth control knobs (Illus. 4-18–4-20). It is frustrating trying to adjust plunge routers in router tables without these features. Some companies that make router models with trigger switches on the handle also make the very same model with flip switches on the motor unit housing. The latter model is, of course, usually more effective and convenient to use in stationary tables and in similar circumstances.

Depth-control knobs on plunge routers should be standard equipment. At least, however, they are now available as accessories to fit major routers, and are definitely recommended to be used on routers designed for inverted stationary use, as in a router table (Illus. 4-20).

Plunge Bases

Porter-Cable produces a plunge base (Illus. 4-21 and 4-22) that permits all 3½″ diameter motor units from

Illus. 4-20. Depth-control knobs for most routers are now available as accessories and are highly recommended for plunge routers inverted in stationary tables when used.

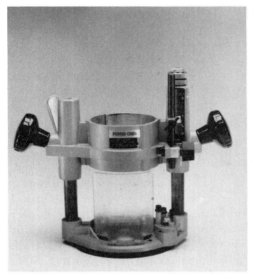

Illus. 4-21 (above left). Porter-Cable's plunge base accepts 3½ diameter motor units from fixed-base routers and features a 2½" plunging depth capability. Illus. 4-22 (above right). Porter-Cable's plunge base accepts many of the company's own motor units and motor units from other companies, including some from Bosch and Black & Decker.

fixed-base routers to be used with it, thus achieving plunging versatility. This product will receive most Porter-Cable and the previous Rockwell motors, five different Bosch router motors, and one Black & Decker motor. If you have a fixed-base router with a motor that fits, this item is worth considering. Sears had a similar

Illus. 4-23. The Sears plunge base has a maximum plunge-depth capacity of only 1".

plunge-base unit (Illus. 4-23) in its 1990–91 catalogue, but I haven't seen it advertised since then.

Overview of Plunge Routers

Porter-Cable currently sells three plunge routers. The smallest is the model 693, 1½ hp router shown in Illus. 4-22 and 4-24. The company's other two plunge routers are 3¼ hp machines. One is a single-speed, 21,000 rpm router (Illus. 4-25). The other is a 5-speed, 10,000–21,000 rpm router.

Skil's largest router to date is a 2¼ hp, 2,300 rpm plunger router (Illus. 4-26). This router and Skil's smaller plunge router both feature spindle locks and convenient wrench snap-in storage on top of the router (Illus. 4-27).

Black & Decker's offers two consumer plunge routers (Illus. 4-28 and 4-29). Its model 7615 router is a 1½ hp unit, which features a built-in work light and storage space for its wrench in its base. It has a ¼" collet, a 25,000 rpm motor, and a trigger-type switch. The model 6200 Quantum plunge router is the newest router. It is a 1¼ hp unit which appears to have features similar to the Elu and DeWalt lightweight routers. The *RYOBI* model 150 router (Illus. 4-30) and the AEG router are two other ¼" routers of approximately the same light weight, but they have considerable differences in price.

Illus. 4-24. A comparison of the design features of popular plunge routers in the mid-size to heavy-duty range. On the top row, from left to right are Makita, Hitachi, and Freud routers. On the bottom, from left to right, are RYOBI, Jepson, and Porter-Cable routers.

RYOBI has two series of bigger routers: a model 500 series (Illus. 4-31), and a model 600 series (Illus. 4-32). There are major differences in the base shapes and handle designs of these two series.

Illus. 4-25. This Porter-Cable single-speed production plunge router looks much like its 5-speed mate. Both are 3¼ hp and weigh over 17 pounds.

Illus. 4-26. Skil's largest router to date is this 2¼ hp professional router.

Illus. 4-27. A convenient storage space for wrenches is located on top of the Skil plunge routers.

Illus. 4-28. Black & Decker's model 7615, light-duty plunge router has a ¼" collet and runs at 25,000 rpm.

Illus. 4-29. Black & Decker's model 6200 Quantum plunge router.

Illus. 4-30. The RYOBI model 150 routers are small and lightweight.

Illus. 4-31. RYOBI's model R-500, 2½ hp router.

Illus. 4-32. RYOBI's 600 series routers, which have 3 hp, are the company's largest routers. Shown here is the model RE-600 router.

The *Elu* and *DeWalt* 1 hp routers are essentially the same in overall design and features (Illus. 4-33 and 4-34). DeWalt's biggest router is a 15 amp, 3 hp router that incorporates many features of the Elu 2½ hp router (Illus. 4-35).

Makita sells a light-duty, 1¼ hp router and two 3 hp models that are exactly the same except that one comes with a round base (Illus. 4-36) and one with a square base. The 3 hp model with the square base, introduced

Illus. 4-35. DeWalt's biggest router is this 3 hp, variable-speed router that has many features of the Elu 2½ hp router shown in Illus. 4-1.

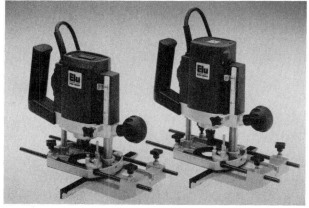

Illus. 4-33. Elu's small 1 hp routers. On the left is the model 3303 single-speed router. On the right is the model 3304 variable-speed router.

Illus. 4-34. The DeWalt routers are copies of the Elu routers. Shown here is a DeWalt 1 hp router.

Illus. 4-36. This Makita 14 amp, 3 hp, 23,000 rpm, single-speed router has remained virtually unchanged for over ten years.

to the United States market over 10 years ago, was the first plunge router import to create serious interest. Its design is still very much the same. It has a 23,000 rpm, 3 hp motor. Makita currently does not offer a variable speed router, which limits its practicality for routing with the new big bits such as the panel raisers, which must turn at much slower rpm.

Illus. 4-37 shows the biggest *Bosch* router: the model 16EVS, 3¼ hp router, which has a fine-adjustment knob and a lock on the trigger switch that makes it suitable for router table as well as hand-held use.

Buying a Router

If you intend to do much work involving following templates, patterns, etc., pay careful attention to the quality and quantity of template guides available for a specific router. (Refer to Chapter 13 for more on template guides and template routing).

Before actually purchasing any router, be sure to analyze these features according to your specific needs: its size and weight; the size of its collet, its horsepower and speed (single, multiple, or variable); its shaft lock, switches and plunge-locking systems; and any other feature that is important to you. To select the router best suited for your purposes, you will first have to determine whether you will use your router more for portable hand-held work or mounted in a router table. Make your choice only after you've had the opportunity to hold, operate, and actually use the router.

Illus. 4-37. The 3¼ hp, electronic, variable-speed router is the biggest router Bosch sells.

Router Bits

5
Basic Information

Since it is the bit that actually produces the cut surface in all routing operations (Illus. 5-1 and 5-2), serious consideration should be given to various factors relating to bit performance (Illus. 5-3). You should be familiar with the various types of bit (Illus. 5-4 and 5-5), know common terminology, which bits to use for certain jobs, the range of quality available when purchasing bits, and how to maintain and sharpen bits.

In my two other router books, *Router Basics* and *Router Jigs & Techniques*, I devoted many pages to the continuously evolving router bit technologies. The combined research and development by governmental agencies and private industry have provided woodworkers with more efficient bits that have better cutting geometries. This, in

Illus. 5-1. One of the ultimate tests of any wood-cutting process is to make cross-grain cuts on softwood such as this butternut, whose weak fibres are usually crushed rather than severed cleanly. These end-grain cuts, all from the same board, were made, from top to bottom, with a router, a band saw, a table saw, and a handsaw.

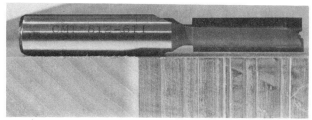

Illus. 5-2. Hardwood end-grain cut with a router is incredibly clean and smooth.

Illus. 5-3. Here a craftsman is making a cut freehand through tough ¾" plywood in a single pass. This requires a router of substantial horsepower and a bit strong and durable enough to make the cut.

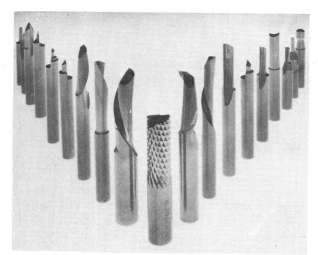

Illus. 5-4. Here is just a sampling of solid-carbide bits of smaller cutting diameters designed for specific routing jobs. The diamond-cut one in the foreground is made especially for routing fibreglass, epoxy, and other hard, nonmetallic abrasive materials.

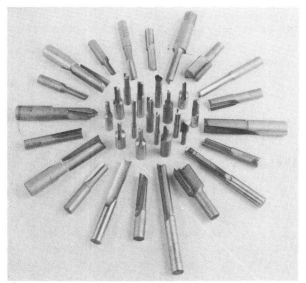

Illus. 5-5. A selection of various bit configurations, all of a straight-flute design for making straight cuts. Note the various cutting lengths and cutting diameters.

turn, permits the development of many bits with new cutting profiles and allows the router to be used for new tasks. Although some material from my other books is repeated here, I encourage reading the above titles for broad-range information concerning router bits.

In this chapter, router bits are discussed in general and information is provided to help you evaluate bits effectively. Though it is impossible to illustrate every style, size and shape of bit available (Illus. 5-6), I have included a generous representation of the bits available. There are many specialty bits designed to make just one specific kind of cut or joint. These types of bits, such as the lock mitre, stile-and-rail, and raised-panel bits, are illustrated later in the book. Specialized bits that have potential secondary uses are illustrated in this chapter (Illus. 5-7).

The amount of work actually accomplished with a good router bit is almost too overwhelming to comprehend. Every second it's rotating, a router bit makes an incredible number of cuts. One individual cutting edge of a typical router bit, for example, if driven at 28,000 rpm makes 467 cuts every second. A bit with two cutting edges would give over 930 cuts per second, or 55,800 identical cuts every minute.

Although most bits are designed to do side cutting in most jobs, many times they have to work as boring tools as well. They have to make their own entry hole into the routing area, so both the end and side edges of the bit are put to work (Illus. 5-3 and 5-7). The bits must also throw out the waste chips from the cut.

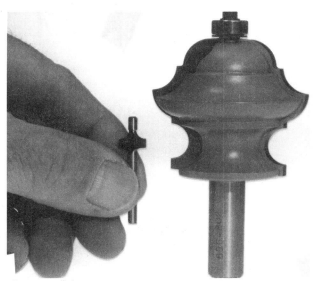

Illus. 5-6. Router bits are available in an endless variety of cutting types and profiles, ranging from the very small ⅛″ shank bit with a single roundover profile, shown on the left, to the large, multiple-profile, ½″ shank bit that cuts dozens of different moulding shapes, shown on the right. This large bit, manufactured by CMT Tools, also features a new anti-kickback design and Teflon® coating.

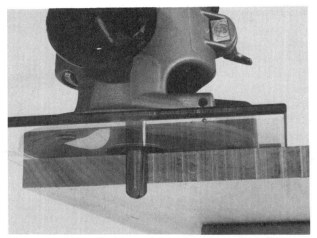

Illus. 5-7. All straight bits and this type of roundnose bit wear mostly ¼″ to ½″ from their tips or ends. Here the roundnose bit is being used for trimming, an efficient secondary use for such bits.

Illus. 5-8. A good, inexpensive set of bits for beginners. These bits are all one-piece, high-speed-steel bits with ¼″ shanks. From left to right, they are a ¼″ straight bit, a ½″ core box bit, a ¼″ radius cove bit, a ⅜″ radius roundover bit, and a ⅜″ rabbet bit.

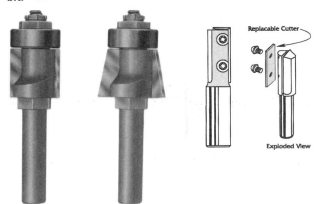

Illus. 5-9. Left: Two throwaway trimmer bits by Paso Robles Carbide (MSM Carbide Tooling and Design). They have replaceable 2-wing cutters mounted on hex-drive arbors that are discarded when dull. Right: A plunge-cut bit with replaceable cutting edges by Trend Routing Technology. The blade has two cutting edges. It is rotated 180 degrees for new sharpness before disposal.

Thus, in terms of the work it does, a good router bit is indeed one very special bargain. Still, many woodworkers expect bits to accomplish even more than they actually do. They feel that a bit should perform (often under impossible conditions) indefinitely with minimal care. This is the wrong approach to take. Router bits should be maintained carefully. A bit that has not been properly sharpened or cared for will require you to use more physical exertion on the router. Also, the resulting cut will not measure up to the standards of fine workmanship.

Cutting Edge Designs and Material Composition

Router bits were first made of special metal called high-speed steel (hss for short) (Illus. 5-8). In the 1950s, tungsten carbide began to be used to make bits. Small pieces of this hard material are brazed to make cutting edges on a tool-steel base material. These are called carbide-tipped bits (Illus. 5-9 and 5-10). They have an edge life 15 to 25 times longer than conventional hss bits.

The best bit for the job at hand is not necessarily the most expensive bit (Illus. 5-8). "Throwaway" bits (Illus. 5-9) and bits with replaceable cutting edges are becoming popular in industry, but as a general rule they are not as yet practical for the home shop woodworker. Generally, what is referred to as "insert tooling" are bits with replaceable cutting edges that are mechanically clamped to the tool body with screws.

There are high-quality carbide bits and cheaper carbides. Better carbide bits have a greater quantity of thicker tips, allowing for more sharpenings. A good production-quality, carbide-tipped bit can be reground up to 15 times. Cheaper carbide bits must be discarded or retipped after being reground four to six times. Some bits, usually smaller-sized ones, are made of solid carbide (Illus. 5-11). Most serious woodworkers today prefer high-quality carbide, which is the most practical choice for long-term use.

The bits that have the longest-wearing cutting edges are diamond bits, which are only used in high-production routing machinery (Illus. 5-13). These bits have a carbide base onto which man-made diamond crystals (poly-

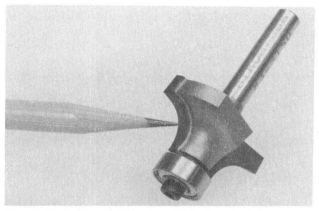

Illus. 5-10. Quality carbide-tipped bits have high-grade carbide. This carbide is fairly thick, so the bit can be resharpened many times.

Illus. 5-11. Left: a ¼" diameter, solid-carbide upcut spiral bit. Note the smooth surface it cuts in pine, and the slight fuzz on the upper surface. Right: a ¼" diameter double-flute, straight carbide-tipped bit and the cut it produces.

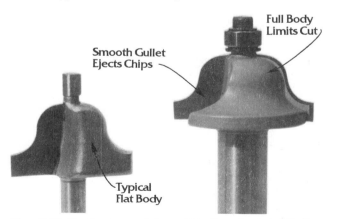

Illus. 5-12. The European-style, carbide-tipped safety bits introduced by CMT Tools, Delta, and Freud recently are highly recommended. They have added material behind their cutting edges that controls the depth of cut of each edge at each revolution. Left: a typical high-speed-steel bit. Right: CMT's new safety bit has an orange Teflon coating that was baked on.

Illus. 5-13. These new industrial diamond bits are very expensive. They have diamond crystals that are bonded to a carbide-base material through a high-pressure, high-temperature process. Diamond bits outperform the best carbide bits, with cutting edges that last up to 300 times longer.

crystalline diamonds) are bonded by a high-pressure, high-temperature process. Diamond router bits are said to retain their sharpness 300 to 400 times longer than carbide bits. They are extremely expensive, ranging up to approximately $1,000 for a typical ½" diameter bit. Still, despite this high price, industry finds them cost-effective in production. There is less sharpening time necessary, the quality of work improves, and feed rates and cutting speeds can be increased, and so factory output also increases.

Experts predict that because diamond is the most wear-resistant material now known, it will be the tool-making material of the future. Diamond tools must be sharpened by experts with facilities that many large woodworking factories do not have. It is not economically feasible for the home craftsman to even consider diamond bits at this time.

Another very promising tool-making material is the "ceramic" developed by Onsrud Cutter. This is a super-tough material with silicon-carbide reinforcing fibres ("whiskers") made from rice hulls. This surprisingly tough material is less expensive than diamond and offers longer tool life than carbide.

Bits with Anti-Kickback Designs

Bits with anti-kickback designs appeared on the European market years ago, but have been slow to reach North America (Illus. 5-12). The bodies of safety bits are much larger than those of conventional bits. The enlarged body is slightly smaller than the overall cutting diameter of the bit, and, therefore, limits the "bite" of the cutter on each revolution, so the chance of kickback is minimal. Bits of this type are strongly recommended for students and beginners, but will prove beneficial to all woodworkers.

The extra body mass of these safety bits offers other advantages than just safety. The extra metal dissipates heat from the cutting edge more quickly, thus keeping the cutting edge sharper for a longer period of time. Heavier safety bits seem to generally ride more smoothly, with less vibration, than lighter ones.

Types of Bits

It is difficult to classify router bits into clearly defined groups or types. In addition to being categorized by the material of their cutting edges, i.e., high-speed-steel (hss), carbide-tipped (ct), or solid-carbide (sc), they could also be classified as one-piece (Illus. 5-8), multiple-part (Illus. 5-14), nonpiloted (Illus. 5-15 and 5-16) or piloted (Illus. 5-17 and 5-18), straight-cutting (Illus. 5-11 and 5-15), surface-cutting (Illus. 5-15) edge-forming (Illus. 5-18), and specialty bits. There is no definite line separating specific groups.

Illus. 5-19–5-21 identify some of the essential terms commonly associated with router bit specifications and descriptions. Below, I describe certain categories of bits.

Surface- or Groove-Forming Bits

Straight Flute Cutting Bits Illus. 5-19, 5-22, and 5-23 illustrate some straight-cutting and surface-forming

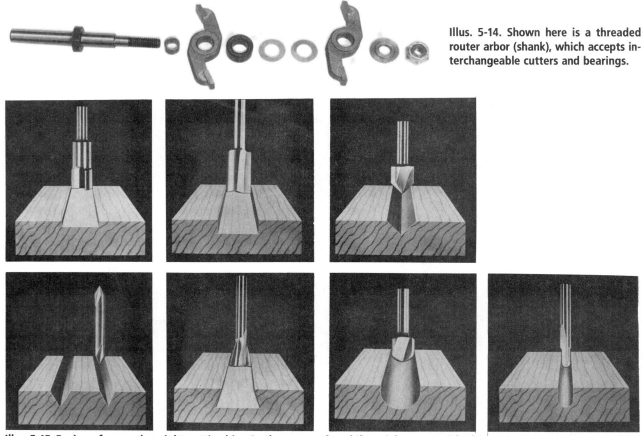

Illus. 5-14. Shown here is a threaded router arbor (shank), which accepts interchangeable cutters and bearings.

Illus. 5-15. Basic surface- and straight-cutting bits. On the top row, from left to right, are straight, hinge mortising, and V-groove bits. Straight bits are used for general routing, cutting dadoes, grooves, circles, and mortises, and most other joint work. Hinge-mortising bits are used for butt-hinge mortising, shallow recessing, and general joint cuts. V-groove bits are used for sign carving, freehand-cutting decorative designs, and with a fence for cutting grooves and chamfering. On the bottom row, from left to right, are 45/60° double V-groove, dovetail, core box, and veining bits. The 45/60° double V-groove bit is used for fine detailing, sign carving, and decorative grooving. The dovetail bit makes all dovetail joints, and can also cut dovetail tenons, tongues, and grooves. The core box and veining bits are used for round-bottom grooving, sign carving, making decorative designs, and freehand work.

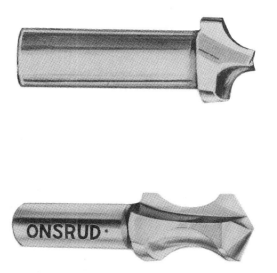

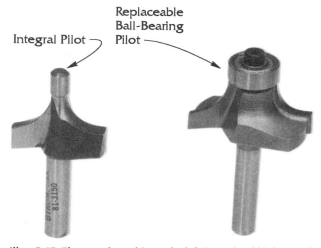

Illus. 5-17. The roundover bit on the left is made of high-speed steel. It has an integral pilot, which means the pilot is part of the bit. The roundover bit on the right is carbide-tipped. It has a ball-bearing pilot that can be removed.

Illus. 5-16. Nonpiloted forming bits. Both bits have boring points.

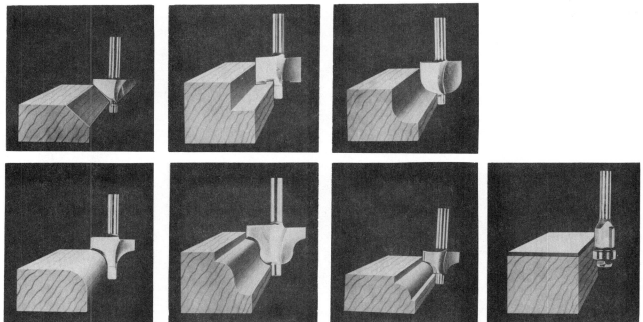

Illus. 5-18. Common edge-forming bits with edge-following or bearing guides. On the top row, from left to right, are chamfer, rabbet, and cove bits. Chamfer bits cut chamfers (usually 45°), and also thin mitres, with the aid of a straight guide. Rabbet bits vary in size. A ³⁄₈″ rabbet bit is most commonly used for cutting to receive backs, bottoms, picture-frame lips, and similar step cuts. Cove bits are used for decorative edge cutting and, with a roundover bit, to make drop-leaf table joints. On the bottom row, from left to right, are corner-rounding (roundover), beading, roman ogee, and flush-trim bits. Corner-rounding bits are used to make decorative edges and drop-leaf table joints, and for general use. Beading bits are used for decorative edge work. Roman ogee bits (with changes in depth) provide various profile cuts. Flush-trim bits make smooth, accurate flush-trim cuts. A ball-bearing guide follows straight and irregular template edges and other guides.

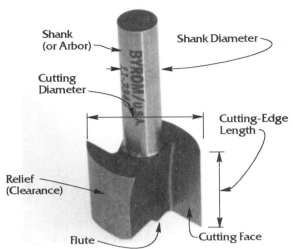

Illus. 5-19. The basic parts of a bit.

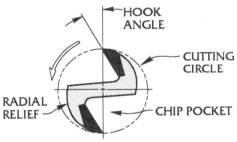

Illus. 5-20. Design provisions for easy cutting include the correct combination of cutting angles (also called hook or rake), side-clearance relief, gullets for chip removal, and perfect rotational balance.

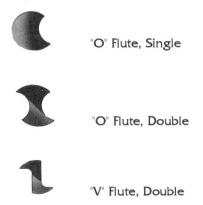

"O" Flute, Single

"O" Flute, Double

"V" Flute, Double

Illus. 5-21. Basic flute designs and cutting edges. These designs have been adapted in most bits to provide them with certain special features, such as more strength, faster chip ejection, or different cutting angles for cutting different materials.

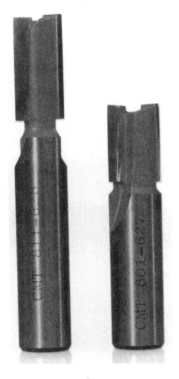

Illus. 5-22. Double-flute carbide-tipped bits. Whenever possible, use the shortest, stiffest bit you can. This reduces chatter, and puts the cutting action closer to the collet and router bearing.

Illus. 5-23. A carbide-tipped, straight V-flute bit. Good bits have bottom clearance bevels and sharpened webs to permit easy plunge entry. Surprisingly, these features cannot be found on all bits.

bits. These bits are used for many routine jobs involving simultaneous bottom- and side-cutting, such as making grooves with straight sides and flat bottoms. Bits in this category are available in a variety of sizes, ranging in cutting diameter from ¹⁄₁₆″ to 2½″ (on ½″ shanks), and in various cutting-edge lengths. Straight-flute cutters are widely used to make grooves, slots, dado, rabbet, lap-box, and mortise-and-tenon joints, dovetail pins, and for other joint-fitting work. These bits are also used to make recessed or through cutouts freehand, or to follow the straight or irregular curves of templates or other patterns with a router guide. Straight bits come in one-, two-, three-, and four-flute configurations, but the one- and two-flute types are the most popular. Illus. 5-24–5-29 show a variety of straight bits that are designed for certain classes of work.

Generally, single-flute bits are stronger and allow for faster feed rates. Two-flute bits make smoother cut surfaces. Straight bits may also be designed with their own integral pilots, or they may have pilots with boring (drilling) points for easy plunging or piercing (Illus. 5-30–5-33). Ball-bearing-guided bits and straight bits with nonboring pilots cannot be used for plunge entry into the work.

Illus. 5-27. An hss double O-flute production bit. O-flutes are stronger than V-flutes, and thus better suited to faster feed rates in solid woods.

Illus. 5-24. Shown here are hss straight single O-flute, 1/4" shank bits. At left is a 1/16" veining bit used for decorative freehand routing, carving, and inlay work. At right is a typical single-flute bit used for cutting softwoods quickly.

Illus. 5-28. A mortising bit developed for hinge butt template routing, but with various other uses. Note the deep center gullet for maximum chip removal.

Illus. 5-29. A double-flute, carbide-tipped chip-breaker bit. This bit is made for the fast cutting of abrasive and dense materials, such as chipboard or particleboard. The edges are designed to break up the chips at high feed rates. The flutes are ground to overlap each other, producing a straight cut.

Illus. 5-25. An hss straight, double-end bit used for general routing, strip inlay, and scroll-cutting. These bits range in cutting diameter from 1/16" to 1/4".

Illus. 5-26. A carbide-tipped, single-flute bit with its end ground for easy plunging.

Illus. 5-30. A group of various piloted straight-cutting bits used for trimming and template- or pattern-guided cutouts.

Illus. 5-31. Panel pilot bits with drill points. At left: a single-flute bit. At right: a double-flute bit. Bits of this type are available in hss or carbide-tipped, in cutting diameters of up to ½″ and cutting edges of up to 1½″.

Ball-Bearing-Guided Straight Bits There are two basic types of ball-bearing-guided straight bit: (1) The standard two-flute, flush-trimming bit with bearings on the bottom; and (2) similar bits, called pattern bits, which have the ball-bearing guide on their shank, above the cutting edges (Illus. 5-34–5-36). Bearings with standard ¼″ and ½″ inside diameters to fit shanks are available in various outside diameters that can be selected to match the bit's cutting diameter if desired. The bearings simply slip over the shanks and are held in place by collars (Illus. 5-35 and 5-36) secured with one or two small setscrews. These two types of bit are widely used to follow templates or patterns mounted above or under the workpiece.

Illus. 5-32. Through-slotting with a panel bit.

Illus. 5-34. Carbide-tipped, flush-trimming bits with ½″ and ¼″ shanks. These bits are generally available with cutting diameters of ½″ and ⅜″.

Illus. 5-33. A stagger-toothed, piloted panel bit with a drill point. This bit is designed to give more chip clearance with higher feed rates. This is an ideal bit for cutting plywood, hardboard, particleboard, and other abrasive material. It is available in various cutting lengths of up to 2½″ and overall lengths of up to 4¾″.

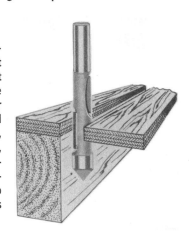

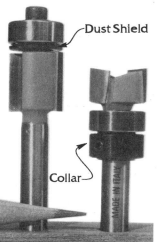

Illus. 5-35. Ball-bearing-guided straight-cutting bits. At left: a flush-trimming bit. At right: a hinge-mortising bit designed to follow a template.

Illus. 5-36. You can make a pattern or template bit from a regular straight bit by mounting your own bearing to the shaft and securing it in place with a collar and setscrew(s).

Shear Bits Shear bits (Illus. 5-37) make slicing cuts because of their inclined cutting edges. They produce smoother cuts with less energy and vibration than the same-size bit with straight cutting edges. Bits with even a slight shear are better than those with no shear. One advantage of shear bits is that they can be honed easily by the owner. Virtually the same honing techniques are employed as when honing straight flute bits because the face or chip sides of shear bits are also flat.

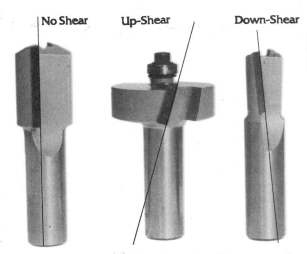

Illus. 5-37. Shear and nonshear bits. Shear bits are usually preferable, because of their slicing cutting action.

Spiral Bits Spiral bits (Illus. 5-38–5-40) continue to gain popularity with woodworkers. Spiral bits are either upcut or downcut bits. Their flutes draw the shaving either up and out of the cut, or force the chips down towards the worktable or through the workpiece. Use downcut spiral bits to make the cleanest cuts, grooves, slots, etc., without surface feathering. Downcut bits tend to pull the router to the work. The disadvantage in using these bits is that they tend to pack the cut or groove with chips more than the upcut bits, which lift chips out of the cutting area (Illus. 5-41).

Single-flute (or -edge) spiral bits allow faster feed rates, but double-flute bits produce smoother cut sur-

Illus. 5-38. An hss single-flute, spiral, upcut bit. In through-cutting, downcut bits are a better choice because they make the chips flow away from the router and operator.

Illus. 5-39. Hss double-flute spirals. At left: a downcut bit. Right: an upcut bit. Double-edged bits make smoother cuts, whereas single-edged bits permit faster feed rates.

Illus. 5-40. This set of hss double-edge spiral bits, by Onsrud Cutter, is highly recommended for a variety of deep-grooving, slotting, and mortising jobs.

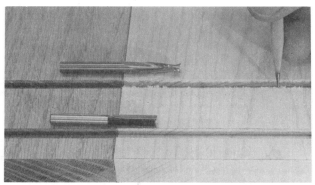

Illus. 5-41. Cutting dadoes in oak and pine with an upcut spiral bit, above, and a straight bit, below. The only noticeable difference in the quality of cut is the surface feathering of the pine with the upcut spiral bit.

faces. The spiral cutting action produced by these bits is somewhat different from cuts made with straight-edged or straight-fluted bits. Unlike straight bits that make one cut, revolve, make another cut, etc., spiral bits make a continuous, uninterrupted cut.

Spiral bits produce less tear-out and a clean, chip-free work surface, and are especially effective for plunging through cutouts, trimming (Illus. 5-42), and when routing material with one good face. The only disadvantage of using spiral bits is that they transmit slightly more torque back to the operator than do straight-flute bits. This makes routing more difficult for freehand work. It is not as much of a problem when spiral bits are used for guided router jobs as long as the workpiece is securely clamped

Illus. 5-42. Here a ¼" spiral bit is being used to trim and clean up a rough saw cut on a pine board.

or attached to a fixture. Incidentally, spiral bits are the bits to use for cutting aluminum. Spiral bits can be made of high-speed steel, can be carbide-tipped, or can be made of solid carbide. They can have one, two, three, or four flutes (cutting edges).

The "chip-breaker" spiral bit shown in Illus. 5-43 is for routing plywood and other man-made materials containing hard adhesives and similar edge-dulling components. Chip-breaker spiral bits are also available as upcut or downcut bits. Downcut bits are better to use for hand-held routing.

Illus. 5-43. A solid-carbide "chipbreaker" spiral bit is designed to handle the toughest cutting materials, such as plywood, composition boards, and soft non-ferrous metals.

Ball-bearing-guided spiral bits (Illus. 5-44 and 5-46) have many uses in the general woodworker's shop. These bits are among my favorites. I use them for trimming, squaring large panels, and template- and pattern-cutting (Illus. 5-47) to produce pieces of various shapes.

Spirals are good general-use bits for cutting nonferrous metals, softwood, and hardwood. They are ideal for cutting profiles, slotting, and deep-plunge routing. Authorities generally agree that spiral bits will always cut smoother than straight flute bits and stay sharp 10 to 15 times longer. Spiral bits are generally available in cutting diameters of ⅛"–¾".

Illus. 5-44. Ball-bearing-guided spiral bits are available in ¼" and ½" shanks.

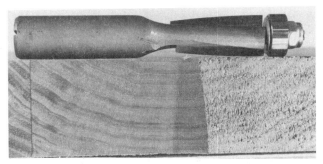

Illus. 5-47. A close-up look at the clean cut produced with this ball-bearing-guided spiral bit made by MSM Carbide Tooling of Oakland, California.

Dovetail Bits Dovetail bits (Illus. 5-48) are among the most popular joint-cutting bits. Dovetail bits all have the same general shape and design, but vary in angle slant from 7 to as much as 18 degrees. They are available with cutting diameters of ¼"–1¼" and with cutting lengths of ⁵⁄₁₆"–1¼". (See Chapter 26 for more information about dovetail bit selection and dovetail joinery.)

Illus. 5-45. A solid-carbide ball-bearing-guided spiral bit with a ½" shank will handle the toughest trimming and pattern-cutting jobs in the average shop.

Illus. 5-48. A 7-degree dovetail bit with a ½" shank.

Roundnose Bits Roundnose bits (Illus. 5-49 and 5-50) are available in high-speed steel, carbide-tipped, and in smaller sizes of solid carbide. They come in straight single- and double-flute styles and in spirals. They are available in cutting diameters of ¹⁄₁₆" to more than 1¼". Smaller-sized roundnose bits are called veining bits. The best veining bit to buy is a solid-carbide bit.

Larger-radius cove-cutting bits (called *guttering bits* in England) are available usually just carbide-tipped and with two flutes (Illus. 5-51).

Another interesting bit which Trend Tools of England calls a *ball groover* (Illus. 5-52) is used for producing decorative round bottom cuts and to hide pipes and cables in structural work. When making the full ball (circular) cut, the pass must be preceded with an initial slot cut that is at least equal to the shank size.

Illus. 5-46. A huge heavy-duty, ball-bearing-guided spiral bit. It has a 1⅛" cutting diameter and a 2½" cutting edge. This bit utilizes a spiral cutter from a power plane and should only be used with the most powerful routers.

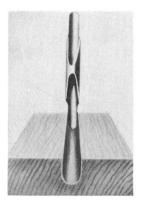

Illus. 5-49. A single-flute veining bit. This bit is used for grooving, making decorative designs free-hand, lettering, and small cove cuts. Veining bits are available in cutting diameters of ⅛″, ³⁄₁₆″, and ¼″, and as hss and solid-carbide bits.

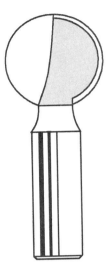

Illus. 5-52. Ball-grooving bits by Trend Routing Technology are available for cutting ½″, 1″, and 1¼″ diameter round grooves.

Illus. 5-50. Teflon-coated roundnose bit by CMT Tools.

Dishing and Tray Bits Dishing and tray bits are among my favorite bits (Illus. 5-53 and 5-54). These are great for routing recesses to make wood boxes and trays, and to remove background material around relief carvings and raised letters of router-cut wooden signs. These bits come in a variety of sizes. A big dishing bit can be as large as 1¼″ in overall diameter and have vertical cutting edges of up to ¾″.

Decorative Surface-Grooving Bits Decorative surface-grooving bits are available in so many cutting configurations and shapes, it's impossible to describe or illustrate every bit available (Illus. 5-55—5-58). These bits all have plunge-entry capability and can also be used to make cuts along the edges of workpieces when appropriate router guides, fences, or shank-mounted ball bearings are used. Today, almost any bit in this category can be purchased or fitted with a shank-mounted bearing of any

Illus. 5-51. Bits like these large cove-cutting bits are available in ¼″ and 2″ shanks and in a variety of radii up to 2″. They are used primarily to make cove mouldings.

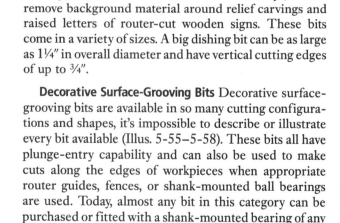

Illus. 5-53. These dishing bits cut a straight vertical and a flat bottom with a rounded corner in one pass. They are useful for trays, boxes, sign-carving, and decorative grooving. They range in cutting diameter from ⁷⁄₁₆″ to 1¼″.

Illus. 5-54. CMT's bits for recess routing of boxes, trays, and wooden signs. Both bits can be fitted with top bearings for pattern duplication work.

Illus. 5-57. A bead-and-fillet combination bit for traditional simulated-panel designs.

Illus. 5-55. This plunge-cutting ogee bit by CMT is ¾″ in overall diameter. It is used primarily to make classical veining cuts in doors and panels, but has many other uses.

Illus. 5-58. A typical panel-raising bit. These bits are available in many different sizes and shapes.

Illus. 5-59. A variety of ball-bearing-guided, surface-cutting bits by Eagle America. They are used to follow patterns/templates of any shape.

Illus. 5-56. An ogee bit for cutting grooves and edge-forming.

diameter desired. Thus, these bits can follow patterns with irregular edges very easily, to make perfect surface-grooving designs (Illus. 5-59).

Point-Cutting and V-Groove Bits Point-cutting and V-groove bits (Illus. 5-60 and 5-61) are also among my favorite types, but good ones are difficult to find. It is

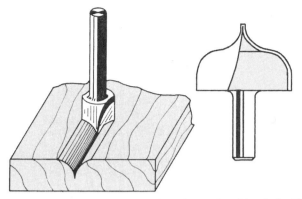

Illus. 5-60. Point-cutting, surface-decorating bits. Left: This point-cutting, quarter-round bit is also widely used for beading. Right: an ogee, panel-moulding bit. Both types of bit are available in various sizes.

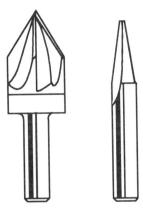

Illus. 5-61. Two V-groove engraving bits by Trend Routing Technology. Both are made of solid carbide. The one on the left has three flutes and cuts very well. Note the slight "flat" on the tip of the bit on the right.

difficult to manufacture and resharpen these bits so that they have sufficient clearance at their point. Consequently, the bit does not always cut cleanly deep in the groove, and often heats up and dulls prematurely. Buy these bits only with the understanding that if they do not make clean cuts, they can be returned for a full refund. Usually, bits of this type work best if they cut a slight "flat" at the intersection of the vertically cut walls (Illus. 5-62).

Illus. 5-62. Round-over bits with a boring point are available in cutting diameters of up to 2″, and with various boring-point diameters.

T-Slot Bits T-slot bits have numerous uses for the creative woodworker (Illus. 5-63–5-65). Bits of this type, unless well designed, often pack their own cuts with shavings and tend to overheat as the shavings are recut. However, when possible, first cut an initial groove to nearly full depth to pre-remove the extra material before making the T-slot cut. This is not possible when making the keyhole slots used to hang various objects.

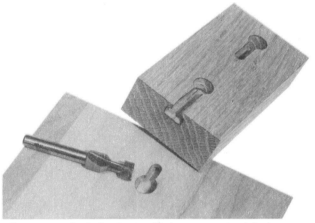

Illus. 5-63. A keyhole-slotting bit by CMT Tools. This bit essentially makes T slots. It is used to make cuts into the backs of picture frames, plaques, and other wall-hung objects. When using this bit, first plunge-cut the entry hole, and then advance the bit. Next, allow the bit to stop rotating, and carefully remove it from the work.

Illus. 5-64. Two T-slot cutters. The one on the left requires a vertical cut to be made prior to its use. The one on the right and in Illus. 5-65 make the complete T-shaped slot cut in one operation.

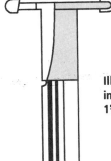

Illus. 5-65. This T-slot cutter by Trend Routing Technology has a cutting diameter of 1⅝″.

Illus. 5-67. A bullnose bit in one pass can cut a full round-over moulding or a bead, depending upon the thickness of the stock cut and the size of the bit. Use this bit only in a router table with a properly adjusted fence or edge guide.

Edge-Forming Bits

In this section, only a few examples of the great number of edge-forming bits available are described and illustrated. These include most popular types of edge-forming bits (Illus. 5-18) and some noteworthy modified bits.

Today, almost all edge-forming bits come with piloted guides. However, bits without pilots can also be used with table-mounted or hand-held routers with edge-guided accessories (Illus. 5-66–5-68). In either situation, straight bits, for example, can be used to cut edge rabbets and full round-end core-box bits used to cut coves. By changing the horizontal or vertical depth of cut, you can use one bit to cut profiles with different shapes and sizes (Illus. 5-69).

Illus. 5-70–5-74 show some common edge-forming bits that are widely used in many project applications.

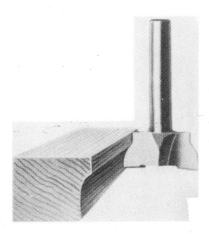

Illus. 5-68. Sash bead-and-cope bits are used for beading the inner sides of window frames.

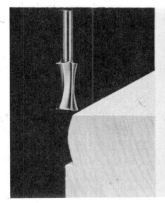
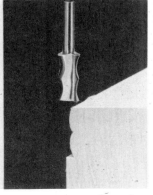

Illus. 5-66. Two edge-forming bits. At left: a large-radius edge-rounding bit. At right: a double-bead edge-forming bit.

Illus. 5-69. Varying the horizontal and/or vertical depth of cut and the position of the workpiece in relation to the cutting bit results in different profiles from just one bit. Here one ogee bit gives four different edge cuts.

Illus. 5-70. These inexpensive edge-forming bits are made of hss, have ¼″ shanks, and cut popular profiles. From left to right, they are chamfer, beading, and ogee bits.

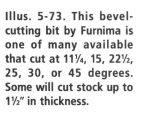

Illus. 5-73. This bevel-cutting bit by Furnima is one of many available that cut at 11¼, 15, 22½, 25, 30, or 45 degrees. Some will cut stock up to 1½″ in thickness.

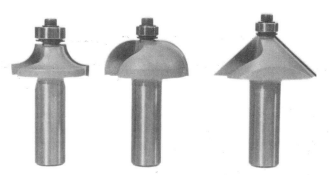

Illus. 5-71. These bits by CMT feature anti-kickback designs, have ½″ shanks, and cut popular profiles. From left to right, they are, round-over, cove, and chamfer bits.

Illus. 5-74. Three-wing slot cutters by CMT Tools.

Slot Cutters

Another type of bit that has many uses is the ball-bearing-guided slot cutter. The best and safest type of ball-bearing-guided slot cutter to use is the 3-wing type shown in Illus. 5-74. Use these cutters to make spline joints, to cut ⁵⁄₃₂″ slots for biscuit joints (Chapter 25) and to cut grooves for T-mouldings. You can control the bit's horizontal depth of cut with the router table fence or an edge guide, or change the diameter of the bit's ball bearings. Three-wing slot cutters generally have cutting depths of ½″. Most slot cutters can be used with slot cutters of different sizes on the same arbor. It is possible to attain a slot-cutting width of ¾″ by stacking the slot cutters from some manufacturers.

Illus. 5-72. Carbide-tipped rabbeting bits. The one on the left is a conventional type made by Amana. The one on the right, made by CMT Tools, features an anti-kickback design.

Rabbet Bits

Rabbet bits (Illus. 5-72 and 5-75) have many uses in the woodworking shop. I routinely cut rabbets to receive cabinet backs and box bottoms, for picture-frame and moulding work, and to make simple, quick joints, including tongues and stub tenons. You will need rabbet bits of various sizes to match various material thicknesses. If you use the new rabbet kit shown in Illus. 5-75, you can cut ⅛″, ¼″, ⁵⁄₁₆″, ⅜″, ⁷⁄₁₆″, and ½″ rabbets of any depth, all from one bit. Simply change bearing sizes. Most conventional, one-size rabbet bits only cut rabbets ⅜″ wide, but by vertically adjusting the bit you can cut the rabbets to any depth.

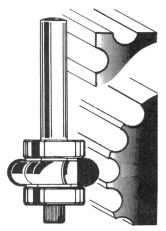

Illus. 5-76. A flute cutter available from Furnima Industrial Carbide in ¼″ and ⅜″ flute diameters.

Illus. 5-75. This rabbeting cutter from CMT Tools has an anti-kickback design, a Teflon coating, and a bearing kit. It cuts six commonly used rabbets, ranging from ⅛″ to ½″.

Illus. 5-77. The ball bearing on this fluting bit permits flute cuts of uniform depth to be made along curved surfaces.

Flute and Bead Bits

Flute and bead bits (Illus. 5-76–5-78) are used for specific tasks, but they have uses in many different types of woodworking. Use bits of this type to groove architectural trim, structural columns, and cabinet and furniture components.

Raised-Panel Bits

Raised-panel bits (Illus. 5-79–5-83) come in two basic types: horizontal- and vertical-cutting. Both types are available in a variety of sizes and profiles. They are used to make architectural wall panels and cabinet doors and drawers. Large, horizontal panel raisers are among the largest bits available; some are almost 3¾″ in diameter. Bits of this size are very dangerous for the inexperienced, and require extreme caution when being used. They

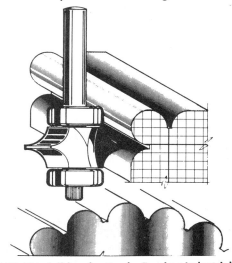

Illus. 5-78. This full-bead cutter by Furnima Industrial Carbide is available with ³⁄₁₆″, ¼″, or ⅜″ radius cutting edges to make one-half or three-quarters of a full bead on round or square table legs, as shown, and similar jobs.

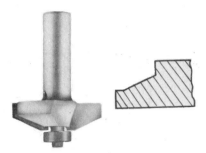

Illus. 5-79. A carbide-tipped panel-raising bit with a ball-bearing pilot.

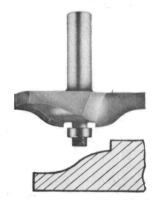

Illus. 5-80. This very large, carbide-tipped panel-raising bit with an ogee fillet design and a ball-bearing guide is available in a cutting diameter that can be as large as 3⅝″.

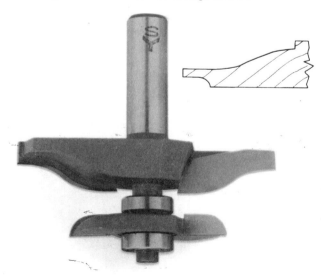

Illus. 5-81. This large panel-raising bit from Cascade Tools, Inc., features an undercutter. The bit is 3⅛″ in diameter. All large panel-raising bits are best and most safely used with speed-controlled routers in sturdy router tables.

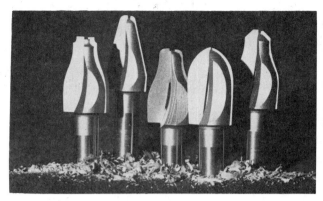

Illus. 5-82. Vertical panel-raising bits from American Wood-craft Tools, formerly Byrom International.

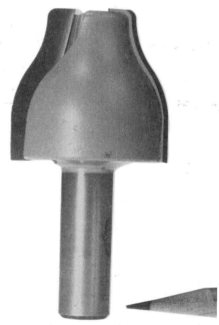

Illus. 5-83. This vertical panel-raising bit from CMT Tools features an anti-kickback design and a Teflon coating. This bit has a 1½″ cutting edge and a 1½″ cutting diameter.

should only be used on a table-mounted router with speed control that is slowed to 8,000 to 10,000 rpm, and successive, shallower passes must be made. Vertical, panel-raising bits (Illus. 5-82 and 5-83) will be preferred by many woodworkers because of their smaller cutting diameters. These bits must also be used in a router table with a high fence. The work is fed on edge over the bit.

Stile-and-Rail Bits

Stile-and-rail bits (Illus. 5-84) are used to make beautifully crafted frames for flat or raised panels. There are several different types, but all are generally of the arbor type and have a ¼″ grooving cutter. Some come in two-bit sets, one bit for cutting stiles and the other bit to machine matching cope-cut rail ends. Most manufacturers provide a choice of stile edge profiles, but the most popular is the ogee.

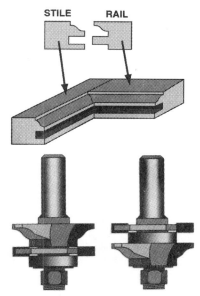

Illus. 5-84. Typical reversible ogee stile-and-rail cutters. Changing the position of the cutter and bearing on the arbor enables the bit to make the appropriate cut. (Photograph courtesy of Cascade Tools.)

Moulding and Decorative Edge-Forming Bits

Moulding and decorative edge-forming bits (Illus. 5-85–5-88). In addition to the typical cove, bead, chamfer, round-over and ogee bits, there are hundreds of other profiles available with which to create that special design on your edge. It is just not possible to illustrate each design option available. Only a few are illustrated here, to give you an idea of the wide range of bits that are actually available.

Multiple-Profile Bits With just one multiple-profile bit, you can make an overwhelming variety of moulding or

Illus. 5-85. Multiple-profile bits from CMT Tools. The one on the left has a ½″ shank, a cutting diameter of 1½″, and a cutting length of 1⅛″. The one on the right has a ½″ shank, a cutting diameter of 2³⁄₁₆″, and a cutting length of 1⅞″.

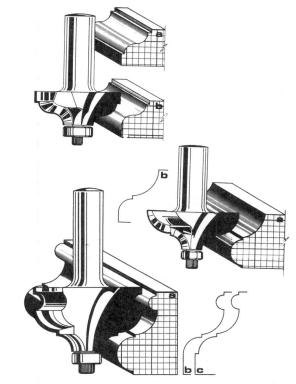

Illus. 5-86. Three edge-forming bits from Furnima Industrial Carbide.

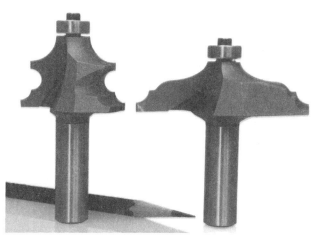

Illus. 5-87. Two of four different bits from Furnima Industrial Carbide designed specifically to produce many profiles of frame stock.

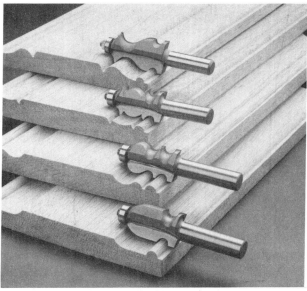

Illus. 5-88. These large-profile face-moulding bits from MLCS Ltd. are used to produce architectural millwork such as baseboards, chair rails, casings, etc.

edge profiles (Illus. 5-85). These bits must also be used in a router table with a fence. Multiple moulding shapes can be created by varying the cutter height, fence position, and number of passes made.

Large-Profile Face-Moulding Bits Sets of large-profile moulding bits are available from most major router bit manufacturers (Illus. 5-88). Bits of this type range in cutting diameter from ¹⁵⁄₁₆″ to 1¼″ and in cutting-edge length from 1⅛″ to 1⅝″. All have ½″ shanks and should be used with routers having at least 2¼ hp for single-pass operations.

Table-Edge Bits Table-edge bits (Illus. 5-89), also called thumbnail bits, make traditional profiles of table edges. They generally have cutting diameters of 2½″. Varying the depth of cut produces different profiles. This bit has more uses than just edge-forming tabletops. Other special-purpose forming and joint-producing bits are illustrated and described in later chapters.

Illus. 5-89. CMT Tool's table-edge bit. Most bits of this type range up to 2½″ in overall diameter.

6
Selecting and Buying Router Bits

Selection Guidelines

With some router bit manufacturers boasting offerings of more than 2,000 bits, it is indeed a problem determining which bits are best for your own woodworking needs.

There are two very important rules that relate directly to good cutting and safety: (1) always select and use a bit with the largest diameter shank that fits your router (Illus. 6-1); and (2) always select a bit with the shortest possible cutting-edge length that will still get the job done. Observing these rules will minimize vibration and promote safer, smoother routing. The following section discusses shank diameter, cutting-edge length, and other factors that will help you determine whether to buy a particular bit.

Illus. 6-1. It's obvious that ½″ diameter shanks (left) are much stiffer and stronger than ¼″ diameter shanks. Both bits shown here have the European-style, anti-kickback design.

Shank Diameter

A ½″ diameter shank bit has four times the cross-sectional area of a ¼″ bit, and obviously that much more stiffness. Bits that have cutting diameters of ¼″ or less should have shanks at least ¼″ in diameter.

Avoid using bushings or reducing collets as shown in Illus. 6-2. These shank adapters allow bits with smaller shank sizes to fit inside large-diameter collets. All routing experts discourage this practice, especially in production situations, since collet reducers tend to increase cutter vibration and do not grip the bit as well as the collet alone.

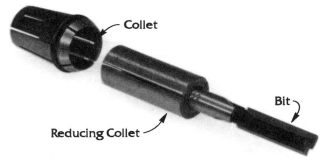

Illus. 6-2. It's always best to match the bit shank size directly to the collet. Avoid using collet reducers (as shown here). They accentuate run-out and vibration and do not provide the bit-holding capability that the collet itself does.

Cutting-Edge Length

Bits with particularly long cutting edges are generally not needed. This is because even when side-cutting, most plunging bits are cutting only on a small portion of material at the lower end of the bit. Even when making a few successive passes to make deep cuts, seldom does a hand-held router exceed a ½″ cutting depth in one pass.

Bits with long cutting edges deflect or bend more under load. This also contributes to vibration and chattering within the cut. Shorter bits are stiffer and, when used, the thrusting forces are closer to the collet and router bearing.

The length of the bit's cutting edge should not exceed three times the diameter. A ¼″ bit should not have a cutting-edge length of more than ¾″. Bits with longer cutting edges are available but some manufacturers do not recommend and guarantee their use, which should tell you something.

HSS Versus Carbide Bits

While it may be true that in the past hss was the only material that could initially be ground to an extremely keen edge, today's improved carbides can be ground to more aggressive rake angles than ever before. Previously, they were considered too brittle. There is not as great a difference in the optimum sharpness of the two materials, especially when you consider the overall edge-wear resistance and much longer life carbide bits have. Carbide bits are generally a better choice, especially when the bit is used to cut a broad range of different wood materials.

High-quality, micrograin-carbide bits produce outstanding finished-cut surfaces in both hardwood and softwood, as well as in other man-made or processed materials, including plywood, medium-density fibreboard, particleboard, plastic laminate, and hard surface materials like Corian®. High-speed steel cutters are still recommended for producing quality finishes when routing natural wood, aluminum, and most plastics. Carbides are definitely the best choice for general jobs, as well as for routing abrasive and cured resin–loaded materials.

Single-Edge (Flute) Versus Multiple-Edge Bits

When speed or a fast feed rate is more important than the smoothness or finished quality of the cut surface, use a single-edge bit. As a general rule, avoid using single-flute bits that are greater than ⅜″ in diameter. They are not symmetrical in design. This causes out-of-balance conditions, bearing wear, and other problems. Use a 2-flute bit when the quality of cut is more important than a fast feed. Three-wing, or flute, bits are helpful when extremely smooth surfaces are required on brittle materials such as plastic laminate. Three-flute cutters can be used effectively at safer, slower spindle speeds and when

low horsepower is a problem. They are also generally safer in large cutting diameters such as are found on better slot cutters and other big bits.

Shear and Spiral Bits

Shear and spiral bits are both great choices and, as a rule, provide the very best of cuts. Both move chips better than straight-flute bits. An edge-forming shear bit provides extra slicing action that is especially valuable when cutting against the grain (Illus. 6-3). Shear and spiral bits will often save labor and money because they have less tendency to tear out wood, and, therefore, ruin edges and waste stock. Shear and spiral bits also generally cut better than other bits when used on routers with less horsepower.

Illus. 6-3. Shear bits (left) are always a good choice for use on smaller collet routers with less horsepower.

A good hss spiral bit is an excellent choice for deep mortising in hardwood. However, the best all-around straight-cutting bits are solid-carbide spiral bits, which are extremely expensive. For example, solid-carbide bits with ½″ shanks cost 6 to 8 times as much as the same-size hss bit. There may be some cost justification, however, because spiral bits will keep their cutting edges 5 to 10 times longer than straight-flute bits. One serious problem is that the average woodworker cannot hone spiral bits, but he can "tune up" straight-flute bits.

Upcut Versus Downcut Bits

Use upcut bits for deep mortising and where chip removal from the cut is essential. Use downcut bits when cutting completely through the workpiece and when you want the chips to flow away from you. Also use downcut bits to eliminate top-surface splintering or feathering. Downcut bits tend to pull the router towards the work surface. This makes them good choices for template- and pattern-cutting, because they help to hold the respective parts in place (Illus. 6-4).

Illus. 6-4. The bit on the left is a short downshear cutting edge and is a better choice for shallow dado and grooving work than the bit in the middle without shear. The roundnose bit shown on the right has an upshear cutting edge that is great for deep cuts.

Chipbreaker and Stagger-Tooth Bits

Chipbreaker and stagger-tooth bits are primarily special-purpose bits used for the fast routing of tough material such as plywood, particleboard, MDF (medium-density fibreboard), and similar man-made abrasive materials. Use these bits where speed is of primary importance and the quality of cut is secondary.

Insert (Throwaway) Bits Versus Brazed Carbide Bits

Bits with insert carbide edges that are mechanically clamped to the tool body have certain advantages over conventional carbide-tipped bits that have their carbides brazed to the tool body. Simply rotating or flipping the edge or installing a new edge creates a high-quality bit immediately without any sharpening or worry about braze (weld) failure. Be advised that most of these bits have single, 0-flute designs, and generally cost 5 to 6 times that of a conventional carbide braze-tipped bit. However, the replaceable blades are of a better grade of carbide.

Selecting Carbide Bits

Most average woodworkers will invest good money in conventional carbide-tipped bits. Our carbide router-bit choices include those imported from the Far East, the Middle East, and Europe, and those made in the United States. There are manufacturers who make impressive claims about bits with impressive prices, and then there are bits whose only claim is actually a low price. Therefore, it is important that you be able to recognize the truly superior bits and weed out the so-called "economy" bits

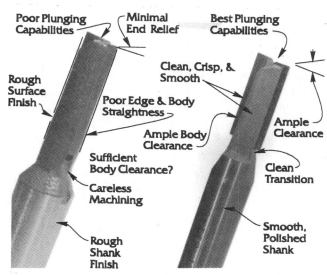

Illus. 6-5. It is possible to compare some bit qualities with the naked eye.

that may put your work, your money, and even your safety at risk.

There are plenty of bits to choose from, but the question remains: Can we spot the good bit when we see it, and before we buy it? The answer to that question is yes, if we know what to look for. (See Illus. 6-5.) A keen eye, a magnifying glass, some common sense, and the following information will help you the next time you go shopping for a bit.

Cutting Edge

Finish Check the finish on the carbide edge. Is it a smooth, polished surface, or can you see irregularities and grinding marks (Illus. 6-6)? Lightly and carefully run

Illus. 6-6. Sufficient end clearance is essential for plunging. Avoid bits with edges having grinding marks this visible.

the edge of your fingernail down the flat surface on the chip-cutting side of the carbide. Does the metal feel rough and wavy? Do the same to the grind on the back side of the edge (Illus. 6-7). Irregular surfaces can be an indication of poor production techniques and careless workmanship. What's more, pitch and gum can build up on rough surfaces and hinder the cutting performance of the bit. Check *each bit* before you buy it; some manufacturers only put a mirror finish on *some* of their bits.

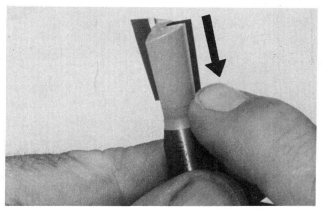

Illus. 6-7. Check the ground surfaces with the end of your fingernail, as shown. You'll be able to detect rough grinds that may not be seen with the naked eye.

Sharpness Testing the sharpness of the cutting edge is also fairly easy. Just rotate the cutting edge gently against the surface of your fingernail (Illus. 6-8). A good edge will shave the nail with very little effort. If the bit won't shave a soft fingernail, do you really want to use it on expensive hardwood?

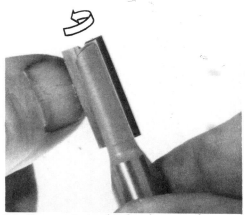

Illus. 6-8. You can test a bit for sharpness by rotating the cutting edge gently against your fingernail.

Radial Relief Look for radial relief on the "back side" of straight cutting edges by looking directly at the end of the cutter (Illus. 6-9). The rounded surface of the carbide provides better support for the cutting edge, which will have less tendency to chip than an edge made with a flat-ground relief surface.

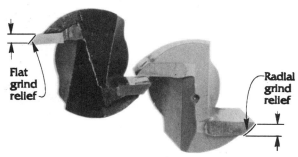

Illus. 6-9. Comparing the "back side" relief grinds and the carbide dimensions of two bits of the same cutting diameter and with other similar specifications.

A bit with a round relief also has another advantage, in that it can be reground many times without reducing its cutting diameter. Manufacturing a bit with radial relief is, obviously, more costly, but results in a more effective bit that lasts much longer.

Straightness Study and compare straight, mortising, and flush-trimming bits from various manufacturers. Observing these relatively simple cutters is a good way to get acquainted with a new line of bits. Sight along their edges to check them for straightness. Producing a straight-edged bit might seem simple, but some manufacturers apparently find it difficult or cannot do it consistently. Let's face it, if a router-bit maker can't produce a straight bit, there probably isn't much point in trying his bits with more elaborate profiles.

Braze Check the braze that holds the carbide to the bit's body. If you notice gaps or voids (Illus. 6-10 and 6-11) in the brazing, beware! A weak or sloppy braze may fail at high speeds, creating a very dangerous situation. The bit should have clean, voidless flutes that do not retard chip ejection in any way (Illus. 6-12).

Quality of the Carbide

The grade of carbide used to make carbide bits may vary substantially from manufacturer to manufacturer. Unfortunately, carbide is very difficult to evaluate visually or to analyze. However, you can make sure you do not buy bits made with inferior carbide if you do the following:

1). Insist on bits with *micrograin* carbide. The hardness of carbide increases as the grain size of the tungsten carbide decreases. Tungsten carbide is actually composed of tiny granules of carbide powder chemically bonded together. The size of the granule is crucial to the performance of the bit. As the bit is used, the carbide gradually wears away and a dull edge is produced. Better carbides can be ground to sharper edges, and the edge will stay sharp up to 10 times longer than cheaper carbides. Cheap, coarse-grained carbide crumbles quickly, resulting in a poor-quality cut. Some manufacturers may even use recycled carbide which is not of uniform quality. Micrograin carbide wears very slowly, so the edge stays sharp longer. Also, less carbide is removed when a fine-grained carbide edge is sharpened, so you'll be able to get more sharpenings out of a micrograin bit. (Illus. 6-13).

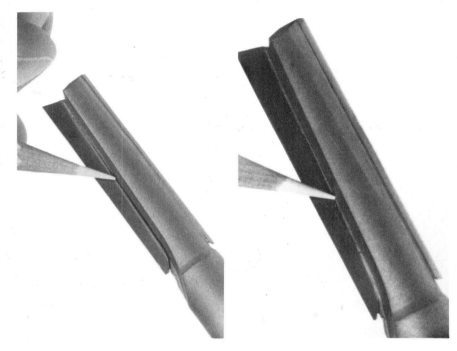

Illus. 6-10. Poor brazing is evident in the left photograph, which shows a void in the weld, or braze, of a bit. The right photograph shows that the other flute on the same bit has a complete braze without any voids.

Illus. 6-11. Reject bits with such obvious voids as shown here. Voids like this can create dangerous situations, and can collect pitch, etc., and retard the ejection of shavings.

Illus. 6-12. Also inspect the back brazing.

  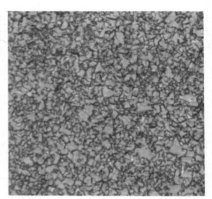

Illus. 6-13. Micrographs of cemented carbide at 1,500 times magnification. From left to right: coarse-grain, medium-grain, and extra-fine-grain (micrograin) carbide. Coarser grades of carbide are cheaper to manufacture and easier to machine, but do not stay sharp as long and are more costly to resharpen. Micrograin carbide is used for all quality bits that have a longer service life and have edges that stay sharp longer. (Micrographs courtesy of Carbide Alloys, Inc.)

2). Compare the size of the carbide cutters on bits from more than one manufacturer (Illus. 6-9). How much more carbide mass does one bit have compared to another? More carbide means more support for the cutting edge and more resharpenings before the bit must be replaced or retipped. Thinner carbide is one indicator of a manufacturer's cost-cutting policy that will eventually affect the performance of the bit.

If you're concerned about the manufacturer's carbide, test one relatively inexpensive bit before buying more from the same manufacturer. A good test would be to run a portion of a brand-new straight bit's cutting edge against particleboard. For example, buy a bit with a 1″ cutting-edge length and cut ½″ thick particleboard. Use a powerful magnifying glass to compare the unused portion of the cutter to the ½″ of carbide that did the work (Illus. 6-14). If you detect wear or a dull edge, you may wish to invest your hard-earned money elsewhere.

Illus. 6-14. Visually inspecting bits with a magnifying glass can reveal imperfections.

Shank and Body Construction

Shank Machining The very first requirement for a good bit is a shank that is machined as precisely as possible to tolerance. If it is slightly undersize, it will not be gripped uniformly around its surface by the router's compression collet. Thus, it will not rotate concentrically to the router shaft, causing vibrations at high speeds. This is not only dangerous, but it puts much strain on the router bearings and wears out the collet, problems that are compounded in a fairly short time.

Look at the shank. Is it smooth, polished, and carefully chamfered at the end? A rough, poorly machined shank can ruin a very good collet the first time it is used (Illus. 6-5). A well-machined shank may be a reflection of the manufacturer's attention to other factors, such as concentricity and dimensional accuracy.

Body Finish Look at the surface of the bit body. Here, as on the carbide surfaces, a rough surface provides a perfect place for pitch and gum build-up (Illus. 6-5).

Cast Versus Machined Bits You may encounter some debate over whether bits that are formed from castings are better to use than bits machined from a solid block of steel. While the cast bits may be perfectly acceptable, it's been my experience that bits that are individually machined from a single bar of steel have well-finished shanks and bodies and are extremely well balanced. You may be more likely to get an air bubble or other impurity in a cheap cast bit than one turned from a block of high-quality steel.

Quality of the Steel The quality of the steel used in a

router bit's shank and body is of great importance. There are differences in the quality of the steel used by various manufacturers.

Some low-cost bits are produced with very soft steel, because it's cheaper and easier to machine. Unfortunately, soft steel is much more likely to bend under heat and stress, leading to early bit replacement (Illus. 6-15). On the other hand, steel which is too hard can be brittle and may break, and the vision of a bit snapping while revolving at 22,000 rpm will give just about any woodworker chills.

Illus. 6-15. Bits can bend in use. The reason may be a combination of circumstances, including soft steel with excessive vibration, run-out, poor collet-to-bit tolerance, worn collets, and overfeeding.

Like the grades of carbide discussed above, the type of steel used in bits is pretty hard to visually evaluate. However, there are certain things you can do to ensure that you buy bits with good-quality steel. First, stick to reputable, established suppliers and manufacturers. Second, read the manufacturer's literature or talk to the salespeople. Can they tell you what type of steel they use and why they use it? Don't expect an in-depth discussion of all the properties of various alloys, but you should be able to glean some basic facts. If toolmakers and their representatives don't seem to know anything about the steel in their own bits, it may be appropriate to ask yourself the following question: If they're proud of the steel they use, why can't they talk about it?

Bit Design and Geometry

Anti-Kickback Design Bits with anti-kickback, European safety designs are beneficial to any woodworker, so I strongly suggest you buy bits with this feature (Illus. 6-16). At least three manufacturers—CMT Tools, Delta

Illus. 6-16. Comparing a new European full-bodied safety bit at right to a more dangerous bit with the conventional design.

International, and Freud—are now selling these safety bits in the United States and Canada. It's rumored that other major manufacturers are also considering producing bits with the anti-kickback design, which features an enlarged body that limits the depth of the cutting edges each revolution.

Side Clearance One often overlooked design feature is the clearance between the cutting edges and the bit body (Illus. 6-17). Proper side clearance minimizes friction and burning while leaving enough space for the chips to be ejected.

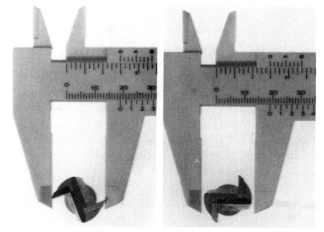

Illus. 6-17. Checking for body clearance. Compare the body diameter of the bit to its cutting (circle) diameter.

Number of Flutes Most of the carbide-tipped bits you encounter probably will have two flutes, which usually is the most logical and cost-effective choice. The surfaces must be smooth and free of any voids that restrict chip ejection.

Shear Angle Shear angle is another important element

of bit design, especially if your goal is a very clean-cut edge. Look at the cutting edge of the bit. If the bit has a shear angle, the cutting edge will appear to slant forward or rearwards in relation to the bit (whether it is upcut or downcut). If the edge looks vertical and aligns with the center axis of the bit, there's no shear angle (Illus. 6-4). Bits with a shear angle slice the wood as they rotate, much as a knife or plane blade slices stock when you hold it at an angle to the workpiece. This action produces a smoother edge than bits with perfectly straight cutters, which tend to chop the wood. Solid-carbide spiral bits are usually preferred, but they are considerably more expensive than straight-line shear bits.

Hook Angle Sometimes called "rake angle," the bit's hook angle determines the angle at which the cutting edge meets the stock. If you wish to check the hook angle, look directly at the end of the bit. Visualize a line extending from the tip of the cutting edge through the middle of the bit. The angle formed by this imaginary line and the front face of the cutting edge is the hook angle.

Bits with higher hook angles tend to feed more aggressively and produce smoother cuts. Bits with low hook angles scrape the wood away rather than slice it. The result is waste that is more like fine dust than soft, thin flakes and chips sliced from the wood.

Plunge Capability To reap all the benefits of the new routers, you need certain bits that are actually designed to plunge. Some very reputable manufacturers make bits with inferior plunge capability. Look at the tip of the bit (Illus. 6-5 and 6-6). There should be clearance between the tips of the carbide and the end of the bit body. However, the cutting edge should not extend excessively or there may not be sufficient support for the carbide. Also determine if the cutting tips are finished with a simple flat surface, or ground with a relief angle for efficient plunge-cutting. The ideal plunge bit also has cutting edges that extend completely across the end of the bit. Even if there is space between the edges it may be possible for the bit to plunge, but you will have to move the bit forward slightly as you lower the router instead of plunging it directly in.

Multi-Piece Bits These are those bit styles designed with interchangeable cutters in various profiles that can be used on one common shank (arbor). Are they a good buy? In terms of avoiding the accumulation of tolerance, no. Since routers turn so fast, you want everything running as concentric as possible to the router spindle.

When you insert a bit into a collet you need some space tolerance or you don't get the bit in. With the interchangeable cutters you add an additional measure to the tolerance totals. The more times you do this, the accumulation of tolerance increases (each successive fit requires some space). It is for precisely this reason that I strongly advise to avoid the use of reducing collets.

Coatings

Several manufacturers now apply coatings to their bits, ranging from a simple paint job to more sophisticated materials such as Teflon® or titanium nitride. Some authorities feel that an extremely thin coating (.0001″) of titanium nitride on the surface of a high-speed steel bit actually increases its hardness. This may be true, but after the bit's first sharpening, the coating will be removed from the chip side of the cutting edge. Eagle America of Chardon, Ohio, claims that titanium nitride provides a sixfold increase in the life of the company's hss spiral bits, but it stopped applying the coating to its carbide-tipped bits. One manufacturer advised me that he stopped coating carbide with titanium nitride because "it caused the carbide to become too brittle." The real reason may be that it's not economically practical to add titanium nitride to carbide tooling for woodworking. CMT Tools and Delta International, on the other hand, have a full line of Teflon®-coated bits. Teflon is commonly used on kitchen utensils because it prevents food from sticking to them. Woodworkers will be equally impressed with its ability to keep resin and pitch from "gumming up" their bits. The slick coating of the body of the bit, and especially in the flutes, expels chips very efficiently.

At least two manufacturers now offer bits with high-visibility finishes. CMT Tools coats its bits with a trademark orange Teflon® coating, and Freud uses a bright-red paint. There is an obvious safety benefit to be gained here. These colors help keep the router user's attention focused on the bit, and that's especially important in view of the ever growing selection of large-diameter cutters.

Other Criteria

Instructions Are instructions provided with the bit? Instructions may not seem important when you're buying a simple straight bit, but you should expect well-written and illustrated instructions when using rail-and-stile bits, raised panel bits, window sash bits, or other complex cutters.

Packaging Does the bit come in a sturdy case that will provide safe, convenient storage in your shop?

Warranty What sort of warranty does the manufacturer provide? Will defective bits be replaced immediately, directly from the manufacturer, or will you have to go through the dealer and distributor? What kinds of damage are excluded from warranty?

Technical Assistance Can you communicate directly with the manufacturer if you have questions about or problems with your bit? Can you make a quick call toll-free?

Price Note that price is last on my list. This is because you should consider the cost of the cutter only after you have factored in all the other criteria discussed in this chapter. Only then can you choose the best bit for the money.

7
Maintaining and Sharpening Bits

Once you've invested in quality bits, there are a few things you should do to keep them in good shape:

1. *Store the bits properly* (Illus. 7-1). If your bits don't come with a protective case or box, drill appropriately sized and spaced holes in a piece of wood so that the bits can stand upright without touching each other. The best-quality router bits will not survive long if they are dropped on the floor or left rubbing against other bits or tools.

Illus. 7-2. Pitch-loaded bits such as this one cannot cut efficiently regardless of how sharp they are.

Illus. 7-1. The best way to store bits is vertically, without touching any other objects.

2. *Keep your bits clean.* Keeping bits clean is as important as keeping them sharp. A bit loaded with pitch and resin deposits on its chip side (Illus. 7-2) or with rusty and pitted flutes is soon a victim of unnecessary heat build-up. This happens because the bit's tool clearances and hook angles become reduced, chip removal is retarded, and the cutting temperature increases. Heat accelerates tool dulling. Resinous deposits can build up surprisingly quickly. When the chips are not ejected, they are recut again and again. As the tool gets hotter, the tars and pitch harden and glaze, more build-up occurs, and the thin, sharp edge dulls prematurely.

Clean your bits intermittently as needed during a job or work period and when you are finished routing. There are commercially available tool-cleaning products, but common gel or spray oven cleaners work as well if not better (Illus. 7-3 and 7-4). Do not try to scrape away hardened deposits with hard tools such as screwdrivers. You may scratch a surface you want to keep smooth or chip the carbide.

3. *Apply protective coatings and lubrication.* Bits kept in damp or humid basements or unheated garages and workshops can quickly get rusty and their polished surfaces pitted. Always remove the bit from the collet when you are finished routing. Apply some dry lubri-

Illus. 7-3. Bits loaded with pitch and resin can be cleaned easily and quickly with spray oven cleaner and an old toothbrush.

Illus. 7-4. Cleaning a bit to almost new condition with household oven cleaner.

Illus. 7-5. Just two of the many products that can be used to lubricate tool surfaces after they have been cleaned (and sharpened). Lubrication prevents rusting and reduces resin accumulation. New, dry lubricants minimize friction by lowering cutting temperatures, thus preventing the tool from becoming dull prematurely.

Illus. 7-6. Lubricating a pilot bearing.

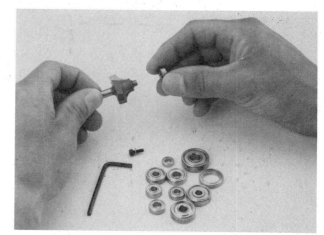

Illus. 7-7. Replace worn, sticky, and frozen bearings with new ones. Also remove them when honing or sharpening the bit. It's prudent to keep some spare bearings of important sizes on hand. (Photo courtesy of Cascade Tools.)

cant to the bit (Illus. 7-5). If the bit has not been used for a long period of time, apply some light oil to it and then wipe it clean and dry before using it. It's also a good idea to lubricate the pilot bearings on the bits regularly with a couple of drops of a synthetic lubricant specially formulated for this purpose (Illus. 7-6). Replace frozen or worn pilots immediately at the first sign of malfunction (Illus. 7-7).

Sharpening Bits

You can determine if a bit is dull in one of four ways. The first is to rotate the edge of the bit against your fingernail (Illus. 6-8, page 88). The second is to visually inspect the bit. Look for a chipped or shiny edge (Illus. 7-8 and 7-9). The third way is to watch how the bit cuts and check the cut surface when routing softwoods (Illus. 7-10). The fourth way is to check the chips. Sawdust, rather than sheared, thin shavings, may also indicate a dull bit.

The very best router bits dull faster than most people realize. Even the best-grade carbide will wear down from continuous cutting of plywood and other resin-filled panel material. Grinding new edges on carbide requires special diamond wheels. Honing and polishing an edge also require diamond wheels. When grinding, honing, or polishing, be sure to attend to all aspects of the bit geometry. Proper relief, clearance and hook angles, along with perfect balance, are as vital to a good bit as is a sharp edge. Most professionals agree that grinding should only be done by qualified experts. That's why some manufacturers offer resharpening services.

Manufacturers actually offer different advice about

Illus. 7-9. A shiny edge indicates a dull bit.

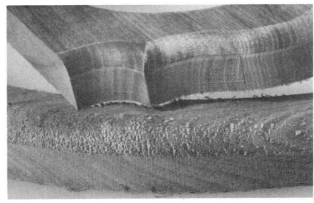

Illus. 7-10. A dull round-over bit was used on these softwood projects. Note the burning, above, and torn fibres, below.

Illus. 7-8. Serious edge-chipping resulting from either cheap carbide or bit abuse. It's obvious that this bit needs professional grinding to restore the edges and balance the tool. Such conditions seriously reduce the overall service life of the bit because so much carbide must be removed.

some aspects of sharpening. Some suggest that the bit's edge can be occasionally touched up by the router user, and others insist that *all* sharpening be performed only by the manufacturer or by professional services. I think it is essential to frequently "touch up" the edges of a bit with a fine hone for the very same reason an expert wood-carver frequently hones his knife on a leather strop. If you send your bits out to a professional every time they need minor sharpening, they will be with the sharpening service more than they will be in your shop. A more practical approach to this whole problem is to have a professional sharpen your bits after you have *lightly* honed yourself several times.

Selecting a Sharpening Service

A good sharpening service is very important, not only because it will ensure good routing, but also because it will protect your valuable cutting tools, which you have made a considerable investment in.

Seek out someone locally whom you can actually visit to see how your bits will be handled. Check the phone book and ask friends for recommendations. A good sharpening service can make bits perform like new, or better. It should be equipped to handle spiral and carbide bits, retip carbide bits, and be able to grind a special bit profile upon request. Should you not find this service locally, then you have to deal with a mail-order tool-sharpening service. You'll find various ones advertising in the woodworking journals.

Inspect your bits when you get them back from a professional sharpening service, and then again after a brief period of use. Use a magnifying glass to inspect their cutting edges and ground surfaces. This will indicate coarse grinds and may even indicate minor chipping along carbide edges, which is unacceptable. Nonvisible conditions of poor grinding may not show up until the cutter is used. If micro-chips appear on the carbide edge of the bit after it has been used a short period of time or it dulls prematurely, the tool may have been excessively heated during grinding. Also, make sure that the service has not ground away more of the bit's material than was needed. This will reduce the number of times a bit can be resharpened. A good grinding service will use wet grinding equipment and put a fine, smooth finish on the ground surfaces.

Inattentive or careless grinding can downgrade a router bit very quickly, and severely damage or even ruin it without your even knowing, unless you make the effort to check your bits (Illus. 7-11).

It's also important that you know when to recognize when the bit is completely worn out. This occurs after repeated grindings, when the carbide has become very thin, the web thickness has been reduced, the bit's basic geometry has been altered, and the bit no longer has an accurate cutting-edge diameter. At this point, it may be wise to discard the bit. Don't be afraid to dispose of a used-up bit.

Sharpening Bits Yourself

Once in a while, a situation may arise where you may want or need to grind a high-speed steel bit yourself. As a general rule, avoid using any router-bit grinding accessory that's powered by a router motor. They grind at too fast a speed. Always clean bits before sharpening or honing, to remove gum and pitch deposits.

One method of grinding straight-flute hss bits that works well is to use a cone-mounted stone in a drill press at speeds not exceeding 1,500 rpm (Illus. 7-12). Grind

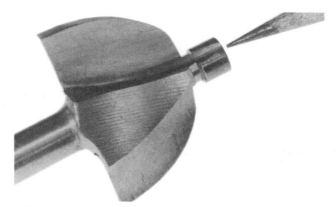

Illus. 7-11. The pilot on this hss bit has been ruined by grinding. Now the pilot has a groove in it that acts like a cutting edge. Actually, the design of the bit, with the pilot so close to the bottom edge, made power-grinding almost impossible.

Illus. 7-12. Freehand-grinding hss bits using a mounted wheel in a drill press.

the bit very lightly. Try to remove an equal amount of material from each wing or flute. Grind only the flat surface on the chip side of the cutting edge, and nothing else. You can also grind bits with flat flutes by working the "flats" by hand on a coarse, flat abrasive stone. Try to maintain the original hook angle, with uniform pressure. Remember to follow with fine honing.

Honing HSS and Carbide Bits All bits with flat flutes, or "flats" in front of their cutting edges, can be honed or touched up quickly and easily with a flat slipstone (Illus. 7-13 and 7-14) or flat diamond hone (Illus. 7-15), whichever is appropriate. Diamond hones are inexpensive and are recommended for anyone serious about routing. You can use them to hone both hss (Illus. 7-16) and carbide bits (Illus. 7-17–7-20) yourself. Be sure to remove any gum and pitch from the bit before honing it.

Because carbide is known to be very hard, you may be inclined to put more work into touch-up honing than is actually necessary. Just a few passes are sufficient. To ensure that the wings, or cutting edges, are honed equally, either count the sharpening strokes or alternate the cutting edges every few strokes. It's surprising how fast a diamond hone can cut away carbide, so stop when done.

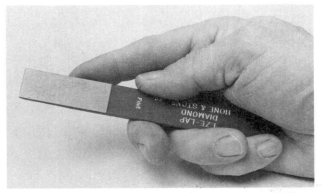

Illus. 7-15. A diamond hone for carbide bits. Use it much like a file.

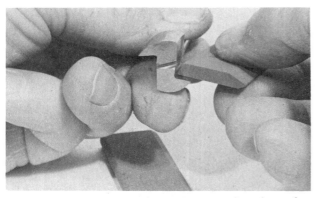

Illus. 7-13. Touch-up honing to rejuvenate the edges of a double V-flute hss bit. Use an aluminum oxide or India abrasive stone.

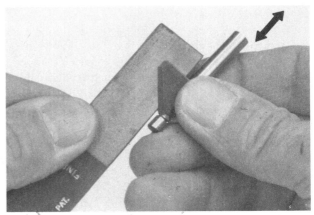

Illus. 7-16. This hss bit is being sharpened on a flat, abrasive diamond hone. Work only the cutting face—the surface towards the flute. Do not sharpen the bevel behind the cutting edge. Also, do not let the pilot touch the abrasive.

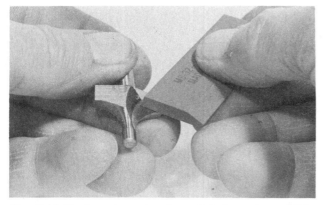

Illus. 7-14. Honing a two-wing, hss, piloted round-over bit is more difficult. Exercise care, to avoid cutting a groove in the pilot. Only very short strokes are possible.

Illus. 7-17. Sharpening a ball-bearing-piloted, carbide-tipped rabbet bit. Note that the bearing is removed.

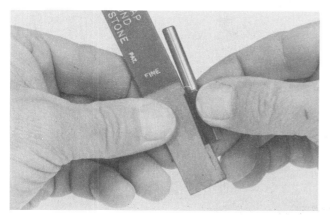

Illus. 7-18. Touching up a carbide-tipped, straight-flute bit.

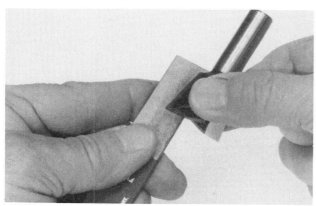

Illus. 7-19. Dovetail bits work hard, because they always must cut to full depth in one pass. Keep them sharp, to minimize overheating and to maximize bit performance.

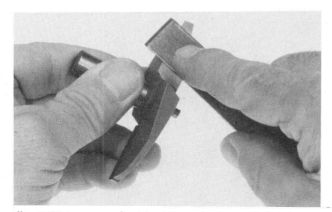

Illus. 7-20. Large panel-raising bits are best operated at slower spindle speeds. This requires slower feeds, shallower cuts, and well-sharpened bits for good, safe work. Here a diamond paddle is being used to sharpen the wings of a big, potentially dangerous bit.

Use a round, diamond honing tool to renew edges on single-flute bits (Illus. 7-21).

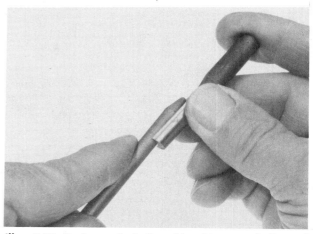

Illus. 7-21. Honing a single-flute bit with a round diamond honing tool.

Power-Honing Furnima Industrial Carbide of Barry's Bay, Ontario, has developed and patented a diamond hone on a ½″ shank to be used with variable-speed routers (Illus. 7-22 and 7-23). You simply install the hone into the collet of your table-mounted router and run it at the router's slowest speed, not exceeding 10,000 rpm. The hone has a fine (600-grit), 2″ diameter diamond face that produces a very good finish. The resin-bonded hone can be trued and cleaned with a special dressing stick (Illus. 7-24). Furnima estimates an effective working life of 2,500 to 3,000 honings with this helpful product.

Illus. 7-22. Furnima's power diamond hone is designed for router table use with variable-speed routers turning only at their *slowest* speed. Note the shank supported on a piece of wood, to minimize tipping of the bit.

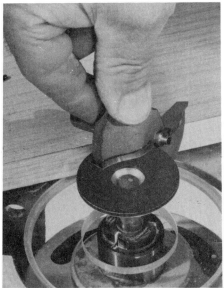 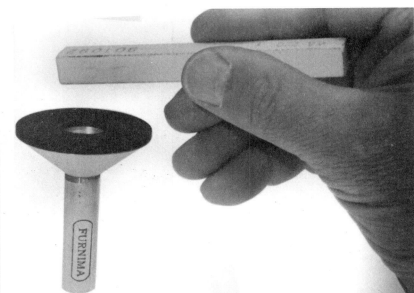

Illus. 7-23 (above left). Sharpening a carbide-tipped, large panel-raising bit on Furnima's power diamond hone in the router table. All sharpening is done only on the inside flat, or chip, side of the bit. Note that the ball bearing is removed. Illus. 7-24 (above right). A dressing hone is also available to clean the diamond face on the Furnima power hone.

Hand-Held Routing Techniques

8
Safety Guidelines

If you seriously value your immediate and long-term physical well-being, you will protect yourself against three hazards associated with routing: eye injury, hearing loss, and lung/respiratory damage (Illus. 8-1). Ways to protect yourself against these hazards are described below.

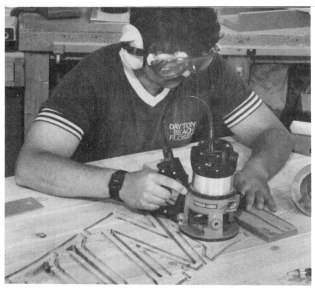

Illus. 8-1. Router-carving a sign involves prolonged periods of operation and considerable material removal. Note that the operator is wearing short sleeves, ear muffs, and goggles, but is not using a dust mask or a dust-collection system.

Eye Protection

Everyone who works with woodworking tools has been repeatedly warned and is well schooled about the ever present hazards of flying chips, sawdust, and knots ejected from power tools. The router presents the same dangers as any other woodworking tool. The chances of a

bit breaking or coming loose during operating may seem relatively slim, but such a situation certainly deserves heeded warnings (Illus. 8-2). There is also the slim possibility that motor screws will vibrate loose and fall into the cutting area. Don't allow yourself to be exposed to these dangerous situations no matter how unlikely that

Illus. 8-2. Bits that have broken during use. The carbide bit shown on the left chipped because it was made of poor-quality carbide and hit a hard object or tough knot. The bit on the right broke because of resin build-up on the tool and in the flutes. The resin build-up occurred because the bit was fed too slowly, causing excessive force and heat. The bit shown on the bottom was destroyed because only a small portion of the bit's shank length was chucked inside the router's collet.

they may occur, or to the obvious hazards of flying chips, without good eye and/or face protection (Illus. 8-3 and 8-4).

Illus. 8-3. Safety glasses with tough polycarbonate lenses and side shields by Cabot Safety Corp.

Illus. 8-4. Safety goggles fit over eyeglasses and give maximum protection.

Hearing Protection

For those who do any amount of routing work, the continuous high-pitched "whir" of the router is far from a comforting sound, and is definitely damaging to hearing. The National Institute of Occupational Safety and Health states that noise-induced hearing loss is one of the nation's 10 most frequent work-related injuries.

A *decibel* (dBA) is the measuring unit for loudness. Sound-protection devices should be worn when sound levels exceed 80 decibels. Under most normal situations, routers operate at over 100 decibels—definitely over the danger threshold.

There are numerous hearing-protection devices available (Illus. 8-5 and 8-6). Basically, they are muffs and various devices that fit into the ear. Muffs are probably effective if the hearing shells are sealed tightly against your head. If they are not, the muffs are less effective.

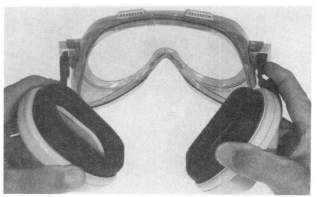

Illus. 8-5. This protection device consists of ear muffs and goggles. It fits over eyeglasses. It can be quickly removed and put on again for short, quick routing jobs. It has spring-in arms that keep the hearing shells snug.

Illus. 8-6. Typical hearing protectors. Shown at the top are earmuffs. These cushions must seal tightly around your ears in order to be fully effective. Expandable foam earplugs (lower left) are the most widely accepted form of hearing protection. Premoulded earplugs (lower right), which come in one size that fits all ears, are also popular.

Long hair and/or eyeglass frames that extend around your ears can break the seal.

Hearing-protection devices (HPD) are given noise reduction ratings (NRR). Some of the new expanding foam earplugs are very effective and give you a custom fit. This type has the highest NRR ratings, and is inexpensive and comfortable (especially for long-term use).

Router manufacturers could do a great service for us if they developed quieter routers. Until they do, the only smart alternative is to wear hearing protection.

Respiratory Protection

More and more articles are being published concerning a potential hazard all woodworkers should be aware of— wood dust. Some kinds of wood dust are more dangerous than others, but all can cause health problems.

The router not only throws out chips and shavings, but it may also generate microscopic dust particles that remain airborne for hours after routing. Like the dust from fine sanding, this type of dust gets into the nasal passages and lungs. Therefore, you should wear protective armor: goggles, earmuffs, and a dust mask (Illus. 8-7).

Illus. 8-8. The author's crude but fairly efficient shop-made vacuum attachment is seldom in the way and collects dust for most routing jobs.

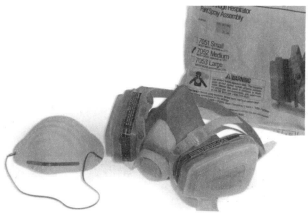

Illus. 8-7. Left: An inexpensive, disposable paper filter dust mask affectionately called the "jock strap" by old pros. Right: The 3-M Company's respirator. This is highly recommended for those who do a lot of routing. Eyeglasses do not steam up when worn under this device.

The ultimate respiratory protection possible is a portable, battery-powered air-purifying system. These systems have air pumps and filters that provide a cool, clean air stream through an enclosed face mask or headpiece. This equipment is recommended for use when you are exposed to certain hazardous jobs, such as when routing old lead-painted surfaces, dangerous formaldehyde-loaded materials, some plastics, etc. If you are ever in doubt about the material being cut, check with qualified authorities for more specific recommendations.

Chip and Dust Extraction

One of the Sears routers has a dust-pickup system built into it. I really do not know how effective it is. I do know that my own vacuum system, though crude, is very effective and powerful (Illus. 8-8). It's connected by a hose

directly to my shop vacuum cleaner. The vacuum cleaner is hung from the shop ceiling over my workbench. With extra lengths of hose, a rope, and a pulley with a counterweight, the router can move freely within approximately an 8′ diameter working circle. The hose is thus always in near-vertical position and out of the way. The slack is taken up automatically by the counterweight. The hose is connected vertically alongside the router with a modified sheetmetal pipe reducer, which is shaped like an inverted funnel. This part was cut to fit as closely as possible into and alongside one of the "webbed," or bridged, openings on the router base where other guides are normally secured or attached.

To make a better air seal, use epoxy paste (from an auto-body repair shop) and putty in the cracks and crevices between the pipe reducer and the router. Before doing this, prepare the surfaces on the router to release the cured epoxy. This is accomplished by first covering all areas with a generous coating of paste wax. As a secondary release measure, I used thin clear-plastic wrap. I also embedded a metal-fastening lug into the

Illus. 8-9. Views of the modification I made to the dust-extraction system I built for my router. On the left can be seen two short rods that were embedded in the epoxy mould, and the flat metal lug with a hole that fits the threaded stud of the handle. The right side shows plastic that was cut out from goggle lens and glued with hot-melt glue over the open bridge of the base.

Illus. 8-10. The attachment is quickly removed simply by unscrewing this one knob.

epoxy so that the vacuum attachment could be held securely in position on the router with one of the base knobs. This entire attachment can be removed simply by unscrewing one knob (Illus. 8-9 and 8-10).

I also found it advantageous to close in the remaining openings under the "bridges" of the base. This was accomplished by using clear-plastic lens material from safety goggles. Cut the plastic to fit and secure it directly to the router bridge and lower base area with a generous bead of hot-melt glue. This lens material must be cleaned occasionally, because the air static tends to attract dust particles onto the material's surface. You can minimize this problem by cleaning the lens material with a plastic cleaner with an antistatic ingredient.

This dust-collecting attachment has made my routing immensely more pleasurable, and it keeps the shop and air much cleaner. It has also improved my routing craftsmanship. If I didn't use the dust-collecting attachment, chips and slivers that would otherwise get partially knocked out of the cut might get stuck in the hole opening in the sub-base and send the router in a wrong direction. Also, the router would tend to crawl over chips, changing the depth of cut when surface-routing. Chips become lodged between the base and the guide, also affecting the straightness of the cut. Another advantage of using the dust-collection system is that I can see the area inside the layout lines because the dust and chips have been sucked away from them.

Two commercially produced dust-collecting accessories that I like from among those currently available are shown in Illus. 8-11 and 8-12. Others have their exiting connections much less vertical; these connections often get in the way in certain routing jobs. Other dust-collecting accessories for routers are described in *Router Jigs and Fixtures*.

Safety Checklist

In addition to protection from the usual hazards of flying chips, noise, and dust, there are a number of additional safety precautions very important to heed faithfully:

1. Read your owner's manual.
2. Make sure that you've selected the proper bit (Illus. 8-13). Check the type, shank strength, cutting length and diameter, and sharpness of the bit.

Illus. 8-11. The dust-extraction kit on this DeWalt router is similar to Elu's. It has a transparent hood, fitting closely around the bit that is held in place by the two rods extending from the edge guide.

Illus. 8-12. This commercially produced dust-collection attachment for some Porter-Cable routers is made by Unique Machine and Tool Co. of Tempe, Arizona.

Illus. 8-13. Here's what happened to a perfectly good hss bit when the user tried to cut a composite material that was too hard for it.

3. Make sure that the router horsepower and speed is appropriately matched to the material, the intended depth of cut, and the size of bit selected (Illus. 8-14 and 8-15).

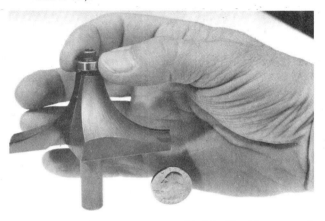

Illus. 8-14. Warning! Do not ever use big bits like this at normal routing speeds or in any hand-held portable routing mode. Use them only and very carefully in heavy routers that are mounted in sturdy router tables, and at slow router speeds and by making successive shallow passes.

Illus. 8-15. A typical 15-amp speed-control accessory suitable for routers. It reduces the rpm of single-speed routers. This makes it safer to use large bits, and matches the speed of the router to bit and material, for optimum cutting. Router motor amperage should not exceed the maximum rating listed on the speed control. Do **not** use these speed controls with routers that have a "soft start" feature. The surge of amperage will burn out the circuits.

4. Always disconnect the power when changing bits, servicing the router, or mounting attachments (Illus. 8-16).

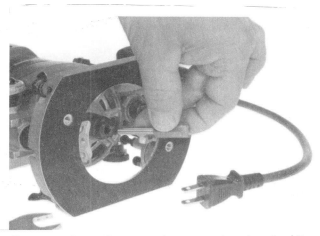

Illus. 8-16. Always disconnect the power when changing bits, servicing the router, and attaching accessories.

Illus. 8-17. Always make sure that the workpieces will be held securely and will not shift or move during routing.

5. Make sure that all bits, attachments, clamps, and locking devices are secured before starting the router.
6. Make sure that the on-off switch is off before connecting the power.
7. Dress properly. Wear eye-, hearing-, and dust-protection devices, and either short-sleeve shirts or long-sleeve shirts with rolled-up cuffs. Wear a shop apron or tight clothing. Make sure that your hair, jewelry, etc., will not become entangled with any moving parts of the router.
8. Keep children and observers at a safe distance.
9. Make certain that all workpieces are securely clamped and will not shift during routing (Illus. 8-17).
10. Always grip the tool tightly, especially when starting up the router, when you have to resist the initial motor torque. Keep both hands on the knobs or handle, or use a foot switch when the job requires a "third hand" (Illus. 8-18).
11. Be especially cautious when working small pieces. Make test cuts in solid unchecked stock of a safe size.
12. Be absolutely certain that the bit is not in contact with the workpiece and that no part of it will strike the wood when you are turning on the power.
13. Develop the habit of switching off the router immediately after you have switched it on. As the motor starts to coast down, use your eyes, ears, and sense of touch to detect any unfamiliar conditions or irregularities.
14. Always shut the power off immediately at the first

Illus. 8-18. A foot switch is not only convenient, but it is also a safety asset when you are using routers with hard-to-reach switches. Foot switches are also helpful for router table and other jig and fixture jobs where complete workpiece control with both hands is required.

sign of any unfamiliar noise or vibrations. Always be aware of the feeling in the handles and the "hum" (that can even be heard through hearing protectors) that indicates the router is operating properly.
15. Do not operate electric routers in moist, wet areas or damp environments.
16. Do not use mounted abrasives, carving burrs, drills, or other nonrouting tools and cutters in routers just because they have shanks that are the same as your router's collet.
17. If the router or work tends to ride upwards and requires extra pressure to feed, turn off the power immediately. This indicates the bit is dull, it is slip-

ping out of the collet, or just that you have selected the wrong bit design.

18. Don't force-feed the router or work in any situation.

19. Do not try to increase the bit's depth of cut by inserting less of the shank in the collet (Illus. 8-2). Most bits should have at least ¾ of their shank length inserted into the collet.

20. Always feed the hand-held router into the work in the correct direction, against the bit rotation. In router table use, feed the workpiece in the direction that is also against the bit rotation. (Refer to Chapter 22 for more on router table safety.)

21. Maintain your router equipment diligently. Replace worn parts, discard worn-out and poor bits, and check the router periodically.

Safety-Related Maintenance

The following maintenance techniques will help ensure good safety practice:

1. Keep all mating areas of the collet and the threads, bit shanks, and spindle clean and free of dust, resin, pitch accumulation, and grit. Periodically remove and disassemble all parts of the collet and give them a good cleaning. Clean them with appropriate solvents or pitch remover, and protect their surfaces with a coating of light machine oil. Resin-filled dust particles can work down the inside of the router into the bottom of the motor arbor spindle through the slits in the collet and become impacted. The only way to remove or loosen the dust particles is with a hard-pointed object such as a scratch awl. Be very careful not to scratch any part of the smooth contact surfaces of the collet or damage any interior threads.

2. If there are any unusual vibrations, first check the bit. It may be bent, chipped, or running off-center. It may also be improperly ground, without adequate relief clearance, or be carrying an excessive chip load due to an excessive feed rate, or just be the wrong bit design for the job.

3. Check bearings frequently. Routers place great strain on their bearings. Replace any bearings immediately if they show any sign of deterioration. If there is still any sign of vibration, disconnect the power and remove the collet and collet lock nut. Turn the motor shaft (spindle) slowly, feeling for rough or irregular rotation. Attempt to push the spindle side to side and then up and down, to detect any movement. There

shouldn't be any movement at all. If there is, it's likely that the ball bearings are rough and should be replaced.

4. Check the collet. Collets do wear, particularly if they are made from lower grades of steel. To determine if it will wear, move a file against it. If it can make a cut against your collet, the collet isn't tempered and it will be subject to premature wear. Vibrations indicate collet wear, which causes bit run-out. Bit run-out means that the bit is not spinning or rotating on its central axis (Illus. 8-19). Get into the habit of inspecting bit shanks as you remove them from the collet. Collet markings on the shank indicate that the collet may be worn.

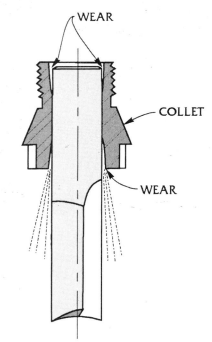

Illus. 8-19. Collet wear and bit run-out.

To make a preliminary check for a worn collet, disconnect the power, insert a fairly long bit, and tighten the collet and lock nut with hand pressure only (without wrenches). Then apply sideways pressure to the extreme lower end of the bit. If the bit appears to move inside the collet, the collet is worn. Get a replacement collet immediately. If vibration still exists, send the router to an expert or back to the factory repair center for appraisal and possible repair or replacement.

5. Friction and stickiness can develop between the metal

motor housing and the metal base of fixed-base routers. Clean these mating parts and apply paste wax or dry lubricant to keep them working smoothly. Also clean and lubricate the posts of plunge routers regularly with dry lubricant or silicone spray.

6. Finally, from time to time blow out the dust from the motor, switch housings, and other areas where it may accumulate. Tighten all screws, and, if necessary, replace worn motor brushes, frayed electrical cords, plugs, and switches.

Here are two maintenance techniques that should save wear and tear on some parts of the router over the long run. First, slip a length of vinyl hose over the handles of collet wrenches so that they do not hit and dent the plunging posts when you are changing bits. Second, apply a coat of paste wax to the router sub-base for smoother feeding with less effort.

9
Basic Routing Procedures and Edge-Forming Techniques

Installing Bits

Always disconnect the power when changing bits. When installing a bit into the router, insert its shank all the way in and then withdraw it $\frac{1}{16}''$ to $\frac{1}{8}''$ before tightening it. If the bit has a fillet or radius between the shank and the cutting edge, be sure *not* to insert any part of this area into the collet. Illus. 9-1–9-4 show techniques for loosening and tightening the collet when removing or installing router bits. Do not overtighten; a firm squeeze or push should be all that's necessary.

Depth of Cut

The depth of cut is the amount of bit protruding below the base (Illus. 9-5). The operator must consider the

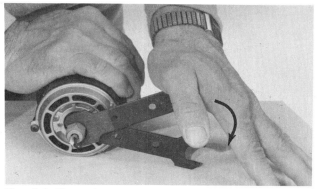

Illus. 9-2. The easiest way to tighten the collet is to use the surface of the workbench for leverage. An open hand, as shown, prevents bruised fingers. Turn the nut clockwise to tighten it.

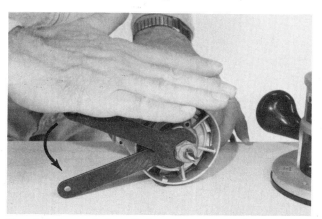

Illus. 9-1. One technique for loosening the collet nut with the workbench used for leverage. The nut should be rotated counterclockwise. Note the operator's open hand.

Illus. 9-3. The two-wrench, one-hand technique for tightening or loosening the collet. Here the collet is being tightened. Exchanging the position of the two wrenches loosens the collet nut.

Illus. 9-4. Changing a bit on a plunge router with a spindle lock. Do not strike the posts with the wrench.

Illus. 9-6. This hard maple was routed with a light-duty router. The final ogee cut at right was made in four successive passes at increasing depths, as shown.

Illus. 9-5. The depth of cut is the amount of bit exposed.

Illus. 9-7. Lower the motor unit until the bit just touches the work surface, and then lock it. This is known as "zeroing out" the router.

width or cutting diameter of the bit, the kind of material being cut, and the horsepower of the router to determine the depth of cut. Multiple passes at shallower settings may be necessary (Illus. 9-6) to arrive at the final depth without placing excessive strain on the bit or router. Estimating the best depth to initially set the router will become easier to do after some routing experience.

To set the depth of cut on a plunge router, you will first have to set the depth scale or stops. First, "zero out" the bit, that is, lower the motor unit so the bit is just lightly touching the work surface (Illus. 9-7). Then adjust the depth stop scale (as required) and clamp the depth-of-

cut stop pole. If it is a deep cut, you can use the rotating turret stops to make the cut in incremental steps with successive passes (Illus. 9-8).

You can expedite making plunge depth-of-cut setups by using shop-made gauge blocks inserted under the stopper pole (Illus. 9-9 and 9-10). There are other methods of speeding up the setup. For example, if you intend to rout a recess for a butt hinge, just place a leaf of the hinge between the stop pole and an appropriate stop on the rotatable turret. This will give you the exact depth-of-cut setting without measuring.

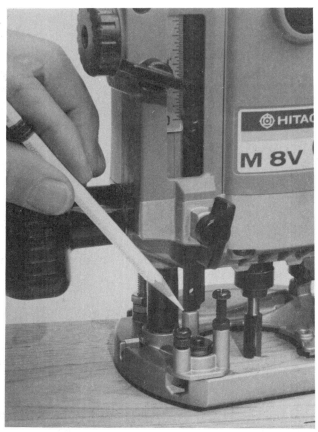

Illus. 9-8. If you are making a deep cut, you can use the stop pole and rotating turret in incremental heights and make multiple passes.

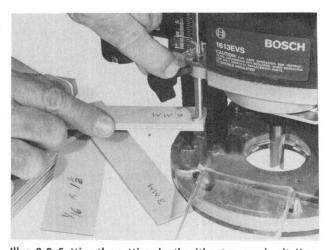

Illus. 9-9. Setting the cutting depth without measuring it. Use a gauge block made from the same material and/or thickness as the material being cut.

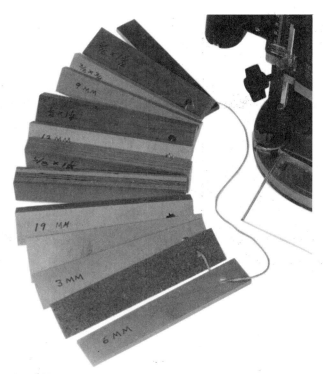

Illus. 9-10. These gauge blocks are made of ordinary wood and are of commonly used sizes.

Throughout the book, I often refer to what I call the "horizontal depth of cut." I use this term to describe the amount of material being cut away when the router is moved horizontally after an initial plunge cut is made; to describe a cut made to widen a previous cut; or to describe a cut made a certain distance inward from an edge.

Practice Cuts

If you are without much routing experience, you should make shallow practice cuts on scrap of the same kind of material you intend to work with. Upon completing a cut, switch off the router. Hold the router steady until the bit coasts to a complete stop. If using a plunge router, retract the bit using the plunge mechanism. If using a fixed-base router, set the router on the workbench. Place the router on its side, with the bit facing away.

Feeds and Speeds

There are two aspects of router work that you must understand and practise correctly in order to handle a router effectively. The first concerns the direction that you feed or push the router into the work (or that you feed the workpiece to a fixed- or table-mounted router). The second is the relationship between the speed (rpm) of the router and the feed rate at which you push or advance the router into the work (or feed the workpiece into a table-mounted router). Both aspects are discussed below.

Feed Direction

The hand-held router must be fed (advanced) in the most opportune direction, which is, for practical purposes, always the direction which counters the force (or torque) of the router. Remember, in hand-held routing, the bit rotates clockwise as viewed from above (Illus. 9-11), and the feed direction should be counterclockwise or against the rotation of the bit (to counter the torque).

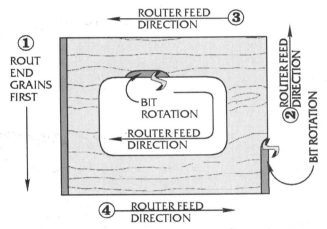

Illus. 9-11. A top-view diagram indicating proper feeding directions for routing around outside and inside edges with a router that's hand-held. Note that the end grains are best cut first.

When edge-routing and facing the routing edge, always feed the router in a left-to-right feed direction (Illus. 9-12). The same concept applies when guiding the router along a straight edge. When you feed the router in the correct direction, the working forces transferred to the bit from the router tend to pull the bit itself into the work.

Illus. 9-11 and 9-13 show techniques for edge-routing all around the perimeter of a board. Here the feed direction, viewed from above the router, is counterclockwise. Incidentally, always rout end grains first. If there is any splintering at the corners, this can be cleaned up when you make the remaining passes, which are cut in the grain direction of the board.

When routing the inside edges of openings (such as

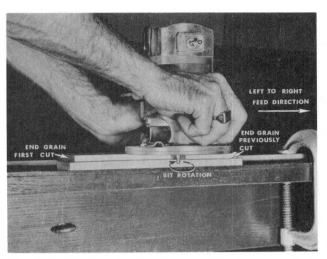

Illus. 9-12. This feed direction is from left to right, or against the direction of bit rotation, as shown.

Illus. 9-13. When you are routing an outside edge completely around a board, feed the router in a counterclockwise direction.

those shown in Illus. 9-11), feed the router clockwise. Essentially, when outside (perimeter) routing is involved, move the router counterclockwise (Illus. 9-13). For inside openings, move the router clockwise.

Pages 259 and 339 contain specific information concerning the safe feeding directions of the workpiece when you are using routers fixed in tables and as overarm routers. Remember, however, that the rule is to always feed the router or workpiece against the bit rotation.

Selecting Router Speeds

Routers with adjustable speed controls can cut more effectively because their cutting speeds can be matched to the diameter of the bit and the kind of material being cut.

Following is a very general reference chart that gives a range of router speeds for various-size bits that *may* work best when you are cutting most wood materials:

Bit Diameter	Maximum Free-Running Speed
Up to 1″ diameter	22,000–24,000 rpm
1¼ to 2″ diameter	18,000 rpm
*2¼ to 2½″ diameter	16,000 rpm
*2½ to 3½″ diameter	8,000–12,000 rpm

*Recommended for use in a router table

Chart 9-1.

Router manufacturers have differing opinions as to what are optimum router speeds for various-size bits when cutting certain materials. This is because there are a number of variables that must be factored in when determining that optimum rpm-to-bit-size ratio. The major variable is horsepower. Large-diameter bits will slow the rpm of routers of less horsepower. Feed rate, cutting depth, general bit characteristics (sharpness, number of flutes, etc.), and the cutting properties (hardness and density) of the material are other variables. It is impractical to provide a router speed chart with specific wood-material cutting recommendations because of all the variables and the fact that routers vary greatly in horsepower. Generally, when it comes to routing plastics, a range of 10,000 to 16,000 rpm is recommended, and for aluminum a range of 8,000 to 16,000 rpm. However, these speeds are also influenced by the size of the bit, depth of cut, and the material's specifications and machining characteristics.

Rate of Feed After you have established the feed direction and determined the best-estimated rpm for the size bit (if using a router with speed options), your next step should be to consider the optimum feeding rate.

The ideal feed rate falls somewhere between too fast a feed rate (which forces the workpiece) and one that is too slow (in which the cut surface becomes burnished) (Illus. 9-14). Knowing exactly the best feed rate at which to advance the router is primarily determined by experience, feel, and the sound of the router. These variables must be factored in: (1) motor horsepower, (2) the machining qualities of the material (hardness, softness, grain direction), (3) bit characteristics (diameter, sharpness, number of flutes), and (4) the established depth of cut.

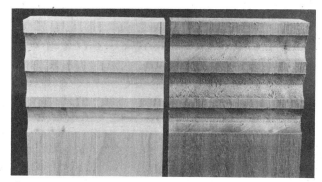

Illus. 9-14. Effects of various cross-grain feed rates in hardwood at left (maple) and softwood at right (willow). The cuts are ³⁄₈″ radius coves made in one pass with a 1½ hp, 23,000 rpm router. The top cut was made at the correct feed rate. The middle cut was made at too fast a feed rate. Note the torn fibres, especially in the willow. The bottom cut was made with the slowest feed. This caused some burning on the maple, but the willow cut is almost totally burnished.

It is possible to use a powerful router with a dull, small-diameter bit and to force a deep cut through hard material fairly rapidly, but the finished cut surface will not be of high quality. Besides, this sort of practice is both foolish and dangerous. A small bit can snap under such stress.

The primary objective in most jobs is to allow the cutting edges of the bit to make clean, sharply severed chips with each rotation. The router must be moved with just the right amount of forward-feeding pressure. When this feed pressure is just right, the bit will not choke on huge chips, or burn and create dust rather than shavings (Illus. 9-15). The kind of waste that your cut produces reveals a lot, not only about your feed rate, but also whether you have selected the right type of bit and a high-quality bit with good cutting geometry. A good cut should make very thin shavings that softly fall to the floor like big snowflakes. If your cut produces fine sawdust, you either are feeding too slowly, or are using a dull bit or one with defective chip ejection and design.

A proper rate of feed is one at which there will be some degree of load on the motor. The sound of a free-running router indicates a proper feed rate. As an example of how to determine proper feed rate, let's assume that we are using a small-diameter, sharp veining bit set to a shallow depth. This bit will put very little load on most routers. The sound of the router does not change much, so it is possible to cut practically as quickly as you are able to advance the router. However, if you were to make a

Illus. 9-15. The waste from the cut should not be dust particles. It should be large, very thin shavings. This indicates that you are using the right feed rate, that the bit is properly *slicing* (not scraping) the wood, and that the bit is ejecting the chips efficiently.

Illus. 9-16. Freehand cuts in Douglas fir, a difficult wood to machine because of its alternating hard and soft growth rings. The bottom shows a wobbly cut, resulting from too slow a rate of feed. The smooth top cut was made with the same bit and router, but with a faster feed rate.

deeper cut or use a larger-diameter bit, you would have to slow the speed of the router (rpm) and the feed rate proportionally. A change in the sound of the router will indicate that it is working harder.

Most beginners (who tend to be cautious) and the majority of all router users probably use a feed rate that is much too slow. The objective of using a good feed rate is to always be in control and maintain your balance. In the majority of hand-fed routing jobs, bits will stay sharp longer, eject chips better, cut cooler and produce smoother surfaces if the feed rates are increased (Illus. 9-16). Machines that are power-fed, such as CNC routers, are programmed to move surprisingly fast, for the most efficient combination of cutting-edge life and cut-surface quality.

The idea of slowing the feed rate to allow the bit to make more cutting revolutions per inch or foot of linear feed, thus making a smoother cut, is true to a certain extent when planing or boring, and for some other wood-cutting operations, but due to the very high edge speeds of routers, higher feed rates are almost always better. This is especially true when routing tough, hard, and abrasive materials, such as MDF, plywood, and particle-board, that dull tools more quickly anyway. These materials should be cut with the fastest feed rates possible, so that the bit cuts fewer chips and is actually in less contact with the material during the cut.

The feed rate should also be continuous and non-

Illus. 9-17. This hard oak shows burns that were caused by hesitations in the rate at which the router was fed. A uniform, uninterrupted feed rate will prevent these markings, which are almost impossible to sand out.

interrupted. A slight hesitation can cause the bit to leave a surface with burned or charred fibres at that spot (Illus. 9-17). Such charring on end grain is especially difficult to remove by sanding or scraping. It's usually better to sharpen the bit and take another shallow, uninterrupted pass with a fairly fast feed.

It should be pointed out that there are a number of routing jobs where it is impractical to maintain a fast or uninterrupted feed rate. Mortising, freehand routing, pattern- and template-routing with sharp curves, and edge-forming inside corners cannot be performed with fast, uninterrupted feeding. Therefore, you may expect some charring at the end of a mortise cut or at an inside corner, when you must momentarily stop or slow down to change directions. With experience, you will learn to "hold up," or pull slightly away, at the most opportune time to minimize charring.

When ending a cut, lift the router and/or retract the bit with the power on. The same technique applies when making grooves or forming cuts in a surface. It is not recommended that the beginner stop the feed and shut off the power while the bit is still in contact with a cut or against the work. The slightest movement of the router could cause the bit to grab the work, kick back the router, and ruin the cut. Similarly, it is not a good practice to start up the router again with the bit reinserted into an existing cut. The starting torque will jerk the router, causing the bit to grab.

Basic Edge-Forming Operations

Of all the many different kinds of cutting jobs that the router can do, making decorative and straight-line edge-joint cuts is one of the most popular. Following are several aspects of making edge cuts that should be discussed.

Holding and Clamping the Workpiece

It is very important the workpiece be held and clamped securely during routing. Often, you can clamp the workpiece so that an edge and the end extend beyond the table (Illus. 9-18 and 9-19).

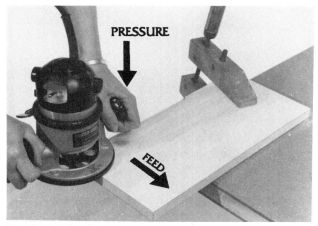

Illus. 9-18. Edge-forming across end grain. Note that the operator must maintain pressure with his left hand throughout the cut, to prevent the router from tipping.

Illus. 9-19. Making a decorative cut along an irregular edge with a piloted cove bit.

Router Pads

Router pads or mats are nonskid-rubber or foam workbench coverings that keep the workpiece from slipping when you are routing, sanding, or doing similar work *on* or *over* the surface of the workbench. These are very useful aids for routing because no clamps or other devices are needed. Unless you are making an exceptionally deep cut, you can use router pads for many edge-forming jobs on stock ¾″ thick and thicker (Illus. 9-20 and 9-21).

Bits to Use When Making Edge-Forming Cuts

Edge-forming cuts are usually made with a piloted bit which follows straight edges (Illus. 9-12, 9-13, and 9-18) or irregular curved edges (Illus. 9-19 and 9-20). Remember that any irregularities along the edges will automatically be transferred to the edge-forming cut because the pilot follows imperfections (bumps, hollows, etc.) as well as clean, smooth edges. Therefore, smooth all edges prior to routing.

Illus. 9-20. Using a router mat. The mat is made of non-skid foam, so it is possible to edge-form with the workpiece held stationary without being clamped or held fast with double-faced tape or other fasteners.

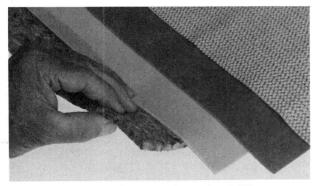

Illus. 9-21. A variety of router mats made of different materials. Thick ones are effective for routing larger pieces. The thin, mesh mat shown on top is best for routing smaller pieces because the workpieces have less tendency to tip under the router.

Ball-bearing-guided bits are the popular choice for edge-forming and the easiest to set up and use. One-piece bits with integral pilots are a little more difficult to use. Excessive horizontal pressure cannot be applied to the pilot during the cut (Illus. 9-22 and 9-23). The torque of the router pulls the bit into the work (cut), which also forces the pilot tightly against the uncut portion of the edge. Consequently, you must exert some physical force to counteract this natural pull. Remember that integral pilots rub against the wood surface they come into contact with and tend to burnish or harden it. This makes sanding and, particularly, subsequent staining more difficult.

Illus. 9-22. Left: a good cut with minimal scoring or burnishing. Right: a poor cut resulting from too much horizontal pressure and too slow a feed rate.

Illus. 9-23. Once charring builds up on the pilot, you must clean it off; otherwise scoring and burnishing of the workpiece will increase.

When edge-routing, always keep the base of the router level and riding flat over the surface. Router accessories are available that help to keep the router "straight up" and minimize the router's natural tendency to tip off the workpiece (Illus. 9-24 and 9-25). Complete edge cuts by lifting the router (or retracting the bit) with the power on.

Edge-Routing Small Pieces

When edge-routing small pieces, you must use a simple, secure way to prevent the work from slipping and minimize the possibility of kickback. There are several ways this can be done; the best method will be determined by the workpiece size and the nature of the cut.

One method is to use a short bead of hot-melt glue or small pieces of double-faced tape to temporarily secure the workpiece directly to the workbench. If a little hot-melt glue is used, the piece will be held securely to the workbench and will not be difficult to release (Illus. 9-26).

Another technique for holding small pieces is to use a nonslip router pad; with this method, clamps, double-

Illus. 9-24. Two routers with base accessories designed to make edge-forming jobs easier. The extra knob located low and away from the router is used to put downward pressure on the inward side of the workpiece, to minimize the tipping tendency which is common to all routers.

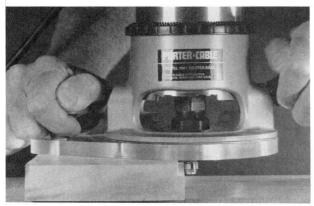

Illus. 9-25. Pat Warner's "offset sub-base" being used to rabbet an edge. This product is available directly from the inventor, at 1427 Kenora Street, Escondido, California 92027.

faced tape, or glue is not necessary (Illus. 9-27 and 9-28). The sub-base with the zero-clearance hole for the bit gives maximum support for the router when feeding it around the edges of small pieces. The thinner style of router pad material is recommended when edge-forming, because the workpieces do not tend to tip over as readily as they do when you use the thicker, foam router pads.

The work-holding fixture shown in Illus. 9-29 and 9-30 is fairly easy to make and is very reliable, especially for production jobs. The fixture shown is used both to shape and round over the inside and outside edges of wooden chain links. In Chapters 13 and 29, I discuss other work-holding techniques that involve the use of abrasives and nail points, as well as inexpensive vacuum clamping techniques.

Illus. 9-26. Routing an inside rabbet cut on this difficult to clamp or hold picture-frame project. It can be held temporarily, but securely, with a spot or two of hot-melt glue or short lengths of double-faced tape, both of which are easily removed.

Illus. 9-27. Edge-routing small pieces on a router mat. The router is fitted with a clear-plastic sub-base, with a very small clearance hole through it for the bit. This base provides the maximum possible support for the router.

Be very careful when making edge cuts that remove most or all of the edge, leaving little or no material for the pilot to ride on. At least 1/16″ of ball-bearing pilots should be in contact with the uncut surface. When using one-piece bits with integral pilots, more than 1/16″ of the pilot should be in contact with the surface, if possible.

Illus. 9-28. Another view of a plastic base. This custom-made base provides excellent router support when you are routing the edges of small workpieces.

Illus. 9-29. This work-holding fixture for making chain links can be applied to similar small work for inside routing. This fixture also serves as a pattern. Here, a ball-bearing-piloted trimming bit is being used to clean up the inside opening, which had most of its material previously drilled out.

Illus. 9-30. The outside routing to round over a chain link. The fixture has an internal plug which is glued to a larger base, which in turn is clamped to the bench.

Routing the Entire Edge

When it's desirable to shape an entire edge, i.e., produce a formed edge that extends from one surface to the opposite face, you must place an auxiliary pattern under the workpiece for the pilot to follow (Illus. 9-31 and 9-32). The pattern must be made to conform exactly to the profile of the workpiece. It can be fastened to the work with clamps, nails, double-faced tape or glue, whichever is appropriate. (Refer to Chapters 13 and 23 for more information about pattern- and template-routing.)

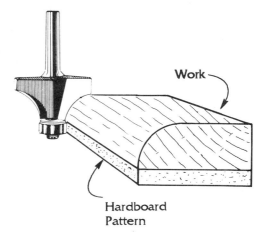

Work

Hardboard
Pattern

Illus. 9-31. Full edge-shaping with a piloted bit with a pattern underneath the workpiece.

Illus. 9-32. Using straightedge material clamped under the workpiece provides a surface for the pilot to ride on when making this full-edge profile with a large round-over bit.

Edge-Cutting Techniques

Illus. 9-33 shows how to make accurate full-edge mitre-joint cuts on thin stock with a chamfer bit. Two ways of enlarging a hole using a rabbeting and a flush-trimming or other straight-cutting bit are shown in Illus. 9-34–9-36. If you have the capability of changing the size of rabbet cuts by changing bearing sizes, this technique allows you to make almost any size hole from ⅜" in diameter upwards in ⅛" increments. Have you ever become frustrated when your sabre saw, scroll saw, or hand-

Illus. 9-34. The first step in enlarging a hole with the router. Here a rabbet bit (with the appropriate-size bearing) is cut to a depth equal to two-thirds or three-fourths the thickness of the stock.

Illus. 9-33. Perfect, full edge or end cuts for 45-degree mitres on thin stock can be made safely and easily with a chamfer bit.

Illus. 9-35. Flip the workpiece over and finish enlarging the hole with a flush-trimming bit. The ball bearing on the flush-trimming bit should follow the part of the hole's wall that was cut with the rabbeting bit.

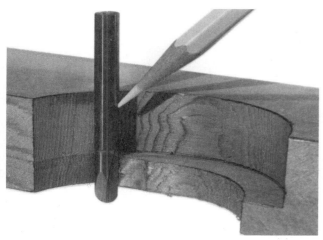

Illus. 9-36. An alternate hole-enlarging idea that also works is to use a bit with straight-cutting side edges and a cutting diameter that equals the bit's shank size. A portion of the shank serves as a pilot and guides the bit around the hole, following the surface that was cut previously with a rabbeting bit.

saw made unsquare or slanted edge surfaces? Your router and a flush-trimming bit can true and smooth these surfaces at the same time (Illus. 9-37 and 9-38).

Router Edge Guides

The router edge guide (Illus. 9-39 and 9-40) is a useful and inexpensive accessory. This attachment allows edge-forming cuts to be made using bits without pilots. The

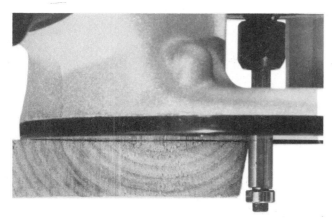

Illus. 9-37. Edges cut with a sabre or scroll saw that are not square or true to the board's surfaces can be easily trued and smoothed with a flush-trimming bit. Simply set the bit to the maximum depth of cut, so that just a portion of the bearing will ride against the workpiece, as shown.

Illus. 9-38. The resulting edge cut is smooth and square with the surface, and only minimal sanding is required to remove cutter/bearing marks. Hardwoods will be cut cleaner than this piece of soft butternut.

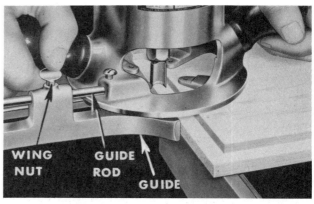

Illus. 9-39. Making an edge-rabbet cut using a pilotless bit and an edge guide.

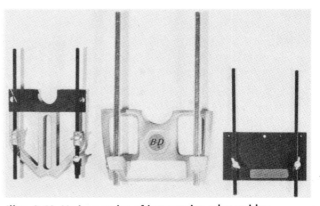

Illus. 9-40. Various styles of inexpensive edge guides.

edge guide also is used to make grooves, dadoes, and decorative cuts parallel to the edge of the work.

Most routers are made with provisions for mounting edge guides and other accessories. Purchase edge guides from the same manufacturer as the router, to ensure a proper fit. Most guides consist of two guide rods and adjustable plates or bars of various configurations. These are all easy to set up and use. Illus. 9-41 and 9-42 show how the guide can be used to rout around the outside edge of a circular shape such as a tabletop.

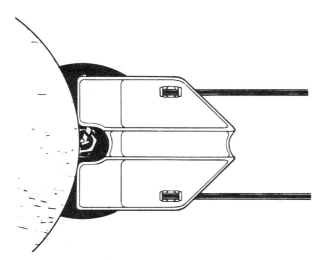

Illus. 9-41. To accurately guide the router along curved edges, use two points of the edge guide in contact with the edge of the stock. This shows the operating position for making a decorative edge around a circular tabletop.

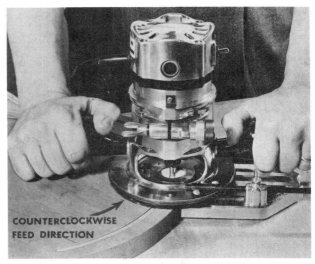

COUNTERCLOCKWISE FEED DIRECTION

Illus. 9-42. Routing a groove parallel to the edge of a circle with the aid of a base guide. Here, the feed direction is counterclockwise, so the torque, or force, of the bit entering uncut wood pulls the guide and router against the work, rather than pushing them away.

10
Straight-Line Routing

This chapter examines some basic techniques employed to guide the router in a straight line. Straight-line cuts are essential to many basic woodworking jobs, such as cutting dadoes, grooves, rabbets, mortises, and many other basic joints (Illus. 10-1). Various techniques for straight-line routing are also involved in trimming and squaring panels, jointing boards, and in making the straight cuts of relief or engraved router-carved letters in signs and inlay work.

Types of Straight-Line Routing

There are two types of straight-line work. The first type involves work in which the router is guided by the edge of the workpiece itself. An example is grooving with the aid of an edge guide following along one edge of the workpiece (Illus. 10-2). In this class of work, the line of cut is always parallel to the edge and will only be as straight as the guiding edge of the workpiece. The other type of straight-line work is those cuts resulting from the router being guided by some other, separate straightedge, clamped or attached to the workpiece. These cuts may or may not be parallel to any edge.

Illus. 10-1. Various straight-line cutting techniques were involved in making the cross-grain dadoes, stopped chamfer, and long, open mortise on the post of this lectern project by Curtis Whittington.

Illus. 10-2. Here straight grooving cuts have been made parallel to the edges of the workpiece with the aid of an edge guide to control the router.

The edge guide on the router acts as a fence. One way to improve its performance is to extend its working surface by attaching a long strip of wood to it (Illus. 10-3 and 10-4). This has the same effect as longer fences and tables do on jointers and table saws, in that it provides more support for easier control. Edge guides permit you to make accurate, repeatable cuts at various depth or width adjustments. This allows you to make deeper cuts with smaller routers or to increase the width of cut using a bit with a smaller cutting diameter (Illus. 10-5 and 10-6).

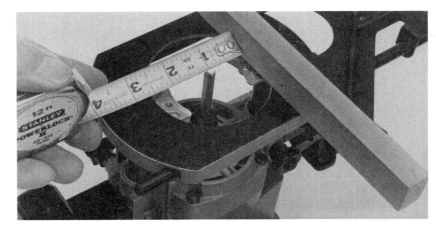

Illus. 10-3. With the strip of wood attached to the edge guide, it is easier to measure to make setup adjustments.

Illus. 10-4. Making this stopped dado is easy with a plunge router and the edge guide. Note that the direction of feed is away from the operator.

Illus. 10-5. A grooving operation. Here a cut is being widened by making a second pass with a narrower bit.

Illus. 10-6. This ½″ × ¾″ end rabbet should be made in several passes with this small router.

Straightedge Guides

Straightedge guides are used to make various joint-fitting cuts, such as those used to make straight grooves, dovetail dadoes, and wide tenons, and to make other types of straight-line surface cuts. A simple straightedge (Illus. 10-7) is used when router edge-guide accessories do not reach, or when the cuts are not parallel to, the work edge. The straightedge is clamped or tacked securely to the surface. It is positioned so that the desired cut can be made with the edge of the router base bearing against it. The straightedge must be held securely in a position that

Illus. 10-7. A simple board straightedge guides the router as it makes identical cove cuts across the edges of these 11 pieces of wood which are clamped together.

is exactly parallel to the intended line of cut. It must also be offset at a distance equal to the measurement from the cutting edge of the bit to the outside edge of the router base (Illus. 10-8). Remember to feed in the direction that is against the rotation of the bit, so that the router pulls its thrust towards the straightedge. Stops can be located appropriately to facilitate making blind cuts and stopped dadoes (Illus. 10-9).

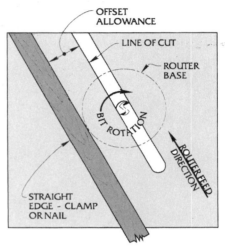

Illus. 10-8. Routing against a straightedge. Note the location of the line of cut and the recommended feed direction.

Illus. 10-9. The through dado cut shown at the top of this photograph was completed with the straightedge clamped in the position shown. The other blind and stopped dado cuts were made with a straightedge and stops clamped appropriately to end the cuts.

A "spacer stick" makes it much faster to locate the offset position of the straightedge than to measure it. The spacer stick is a piece of ⅛″ thick hardboard (or other similar material) in a suitable length, ripped to a width that equals the offset measurement, that is, the distance from the cutting edge of the bit to the edge of the router base (Illus. 10-10–10-14). Rip spacer sticks for bits of various diameters. Then, with a soft-tipped pen, mark the bit size on each of the spacer sticks so that you will know which spacer stick to use in a certain situation (Illus. 10-13).

Illus. 10-12. A straight strip of wood nailed in place guides the router as it makes the straight cuts for the bottoms of letters in this sign work. All the letters will be uniformly sized and straight.

Illus. 10-10. A spacer stick is a length of ⅛″ hardboard ripped to a width that equals the exact distance between the cutting edge of the bit and the router base.

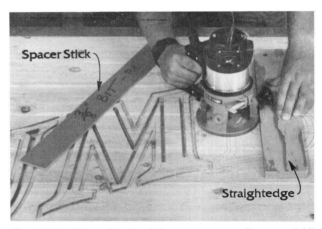

Illus. 10-13. This professional sign carver uses a "spacer stick" and a special straightedge that can be quickly repositioned without clamping to make the straight cuts outlining the letters. Note that this experienced craftsman is guiding the router with one hand.

Illus. 10-11. The spacer stick eliminates the need for repeated measuring when securing the straightedge at its appropriate location.

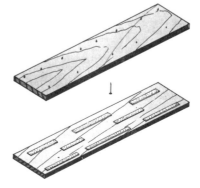

Illus. 10-14. Bottom views of two basic straightedges used for router surface carving. Slightly protruding nails (above) and strips of sandpaper (below) help to hold the straightedges steady without clamping.

T-Square

The T-square (Illus. 10-15 and 10-16) is a very simple device to make, and has many obvious uses for guided router cuts. It is especially handy for making dado and other cross-grain cuts. The lengths of its head and blade can be changed to suit special requirements. However, if its blade exceeds 12″ to 16″ in length, both parts should be made wider, or an extra clamp should be used to secure the blade end opposite the head. (The torque, or rotational force of the router, will bend narrow-bladed T-squares during the cut if they are clamped only at the head.) With the T-square head positioned firmly against the work, splintering at the edges is reduced. When a router cut is made through the head (Illus. 10-16), it makes positioning the T-square to the layout lines easy.

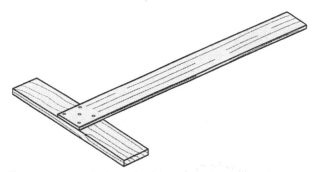

Illus. 10-15. A typical T-square-type straightedge can be made to any convenient size.

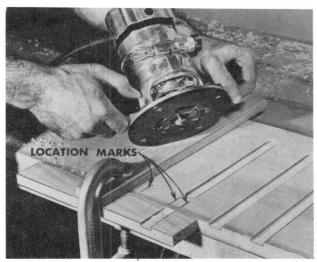

Illus. 10-16. This shop-made routing T-square straightedge is an adaptation of the typical straightedge.

Frame Guide

The frame guide (Illus. 10-17–10-21) is another adaptation of a straightedge. This shop-made device is also simple to construct and extremely useful for accomplishing many routine and unusual routing jobs. This fixture consists of two straightedges (that cradle the router base) and a T-square head set at 90 degrees that can be used to make cuts perpendicular to a trued-up edge.

Illus. 10-17. The frame guide consists of two parallel straightedges spaced (with two cross-members) to a distance exactly equal to the diameter of the router sub-base.

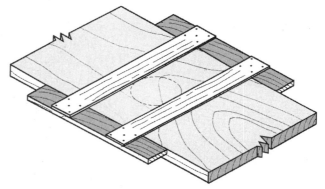

Illus. 10-18. A saddle-type double straightedge is made to fit a specific board or panel width, as shown here.

Illus. 10-19. A frame guide with a T-square-type head set at 90 degrees to the parallel straightedge. When it is used with a spacer stick, quick positioning is easy. Clamp it securely.

Illus. 10-20. Here the frame guide is used to outline the vertical straight cuts of large routed sign letters.

Illus. 10-21. When securely clamped, the frame guide also makes easy work of straight cuts at any oblique angle.

Commercial Straightedges

There are commercial straightedges and many routing guides available that make straight-line router work easier and faster. One very popular design is actually a low-profile, aluminum-extrusion straightedge-and-clamp combination. When used with a spacer stick, it makes setups very quickly (Illus. 10-22 and 10-23). The metal straightedge also makes a good direct-guiding surface for ball-bearing trimming bits and other bits. Illus. 10-24 shows how a flush-trimming bit can be used with the metal straightedge to joint or true the rough edge of a board.

Illus. 10-22. This combination straightedge and clamp with a spacer stick is quickly positioned to the layout lines marked on the edge of the work.

Illus. 10-23. The low profile of the clamping straightedge allows the handles of the router to clear over it.

I've made straightedges that have a thin "spacer stick" offset permanently attached directly underneath them. To make your own, attach a piece of ⅛" thick hardwood or thin plywood to any strip of wood. To use it, make the cut with the router riding on top of the spacer stick. A

Illus. 10-24. You can joint a rough edge straight and true if you use the clamping straightedge with a trimming bit.

commercial version featuring this same concept is shown in Illus. 10-25 and 10-26. That straightedge, called Ply-mate®, is advertised as a guide for a hand-held circular saw, but it also works well for many straight-line routing jobs. The Plymate consists of an aluminum straightedge with a piece of ¼″ replaceable hardboard. The hardboard can be adjusted in or out (Illus. 10-26), and it can be cut to the exact offset distance (bit to fence distance) with the router. The initial pass with the router and the bit you

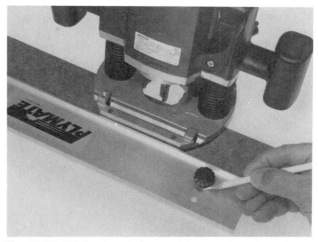

Illus. 10-26. This knob-and-slot provision permits an in-and-out adjustment for changing the correct offset distance of the hardboard spacer material.

intend to use automatically sizes the hardboard spacer material. To make it all work better, attach a small outrigger block to the bottom of the router with double-faced tape (Illus. 10-27 and 10-28). This is cut from the same material thickness and gives the router a stabilized, level ride.

Illus. 10-29 shows the Acculock® triangle fence, which is also made of extruded aluminum. This jig has a 25″ straightedge that can be reset accurately at any point in a 360-degree range. This will guide the router in making specific angular cuts and trimming mitres of any angle.

Illus. 10-25. A commercially made straightedge called Ply-mate® is being used to control the router while it cuts grooves.

Illus. 10-27. Use double-faced tape to attach a small outrigger block to the router that equals the thickness of the offset spacer material.

Illus. 10-28. The small support stabilizes the router.

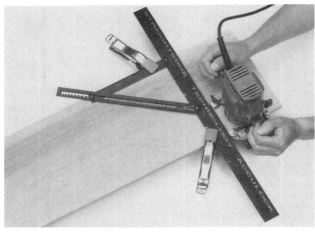

Illus. 10-29. The Acculock® triangle fence is adjustable to any angle. (Photo courtesy of Cascade Tools, Inc.)

Trimming and Squaring

One skill all woodworkers should have is that of consistently making straight and square cuts from any edge in one pass. By "square-basing" your router, using a few easy-to-make jigs, and exploiting the potential of a flush-trimming bit, you will be able to do a variety of such jobs. The following sections examine helpful guides and square-basing techniques.

Square-Trimming Guide

The square-trimming guide shown in Illus. 10-30 and 10-31 is somewhat similar in purpose and function to the T-square guide described previously, but this one has a wider and more rigid straightedge blade. The size given in Illus. 10-30 is only a suggestion; make the guide larger, if necessary, to accommodate your needs. This jig is designed to be used in a few ways. It can be used clamped above the workpiece at the appropriate offset distance required to effect the desired cut (Illus. 10-31–10-33). It can also be used clamped *under* the workpiece without the usual offset allowance. It is set or placed so that the straight edge of the blade is the intended line of cut. The cut is made with a flush-trimming bit (Illus. 10-34 and 10-35).

Use the ball-bearing-guided flush-trimming bit to clean up any protrusions evident after assembling corner joints such as rabbets (Illus. 10-36) and dovetails.

Using a Router with a Square Sub-Base

A square base (Illus. 10-37) has some distinct advantages over any other factory-supplied sub-base. A square base

Illus. 10-30. Details for making an L-shaped trimming and squaring jig.

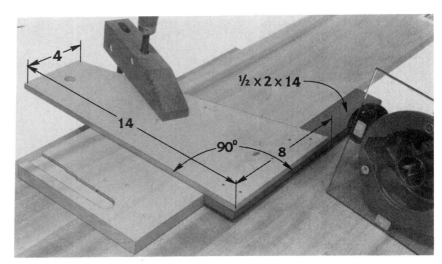

Illus. 10-31. The workbench, workpiece, and square-trimming guide are clamped so that the base of the router will follow along the straightedge. The jig is clamped on top of the workpiece, as shown, and offset a distance from the line of cut that equals the distance from one edge of the base to the cutting edge of the bit. This trimming guide can also be used to cut dadoes and grooves with or without a square-base router.

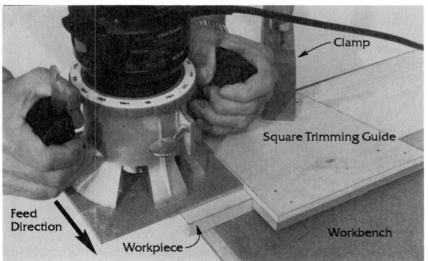

Illus. 10-32. Using a router with a square-trimming guide. Clamp the guide the distance it should be offset from the end of the board, so that you can make the light trimming cut exactly where desired.

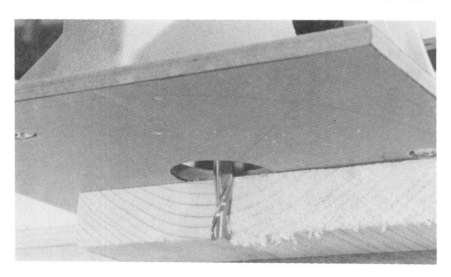

Illus. 10-33. A bottom view showing the cutting action of the trimming operation employed to simultaneously trim to size, square, and smooth the end of a pine board.

Illus. 10-34. The square-trimming guide as set up on the workbench to square an end of a wide, hardwood board with a router and a ball-bearing-guided trimming bit. Note that the jig is clamped between the workpiece and the workbench, and positioned so that the cut will be made beyond the edge of the workbench.

Illus. 10-36. Use a flush-trimming bit to pare down and smooth slight overhangs of assembled corner joints.

Illus. 10-35. Here's a view of the cutting action of the ball-bearing-guided trimming bit when it is used with the square-trimming jig. Note that the board that is to be trimmed square is clamped on top of the jig and to the workbench table. The ball bearing guides the bit, following the straightedge of the jig, as the bit makes a clean, smooth, straight, and square cut.

Illus. 10-37. A router with a shop-made square base can make any of four passes at different distances from the guiding edge without requiring any additional adjustments.

Illus. 10-38. A T-square jig, clamped to the workpiece, and a router with a square base have been used to make a series of dado cuts. As shown here, you can rout dadoes with any one of four different widths without moving the T-square jig or changing the size of the bit. The offset square base of the router makes this possible. Feeding the router into the work with a different edge of the router's square base along the straightedge results in a cut made at a different distance from the T-square guide.

is intentionally made so that the bit does *not* extend through the center of the base, but is instead off-center. Consequently, the square sub-base gives you a choice of two or more different offset distances from the edge of the sub-base to the bit. I always make mine so that I have four different offset distances from each edge to the bit (Illus. 10-38 and 10-39). Depending upon what kinds of cut I need to make, I usually have one edge only about 1/64″ less than one of the other edges so that I can make that last, quick finish-trimming pass against the straightedge without trying to reclamp it 1/64″ from where it was for the first pass.

Illus. 10-40 shows a sub-base I've made from clear polycarbonate plastic. Stickers identify the distance from each edge to the center of the bit.

Illus. 10-40. This plastic, square sub-base has four different distances from each edge to the central axis of the bit. They are: 3½″, 3⅜″ (or minus ⅛″), 3¹⁷/₃₂″ (or plus ¹/₃₂″), and 3³¹/₆₄″ (or minus ¹/₆₄″).

Illus. 10-39. The offset distance is the distance from the cutting edge of the bit to one of the selected edges of the base.

Porter-Cable is the only company I know of that provides anything similar to the above square sub-bases for its routers (Illus. 10-4). The two sides of this sub-base are dimensioned ⅛″ less to the center than the other sides, to provide a ⅛″ trim cut without your having to readjust your straightedge.

Panel-Trimming Jig

Squaring the ends of long, wide boards, squaring large panels, and making mitre cuts across big pieces are among my most difficult sawing jobs. When I need those

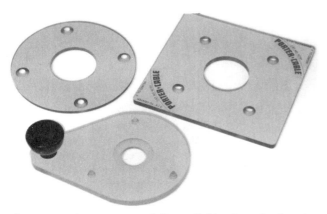

Illus. 10-41. Some commercially available clear-plastic sub-bases. Above: a round sub-base good for freehand work and a square base, both from Porter-Cable. Below: Pat Warner's offset sub-base features two true sides to work off straight edges and a counterbored center hole that accepts Porter Cable's template guides.

kinds of cut to be exact, I saw these pieces slightly oversize and trim them to finished size with the router. It's an extra step, but it ensures perfect results every time. With this method, I can also cut veneered panels without causing tear-out or splintering, which usually occur when I saw them to the finished size.

Illus. 10-42 shows a jig I made for square-trimming and some mitre-trimming. You can make your own any size needed. This jig is just a piece of ⅜″–½″ thick plywood with one sound, straight, and true edge. On top of this edge is an adjustable straightedge that pivots on one carriage bolt. It has another bolt that rides in a radiused slot which can be tightened at any adjustment. Toggle clamps on the straightedge hold the workpiece secure while the piece is being finish-cut with a flush-trimming bit (Illus. 10-43–10-45).

Illus. 10-42. This large panel-trimming jig consists of a plywood base and an adjustable straightedge with some toggle clamps.

Illus. 10-43. Rough-sawing a wide panel end that will be trimmed square and true on the jig with the router.

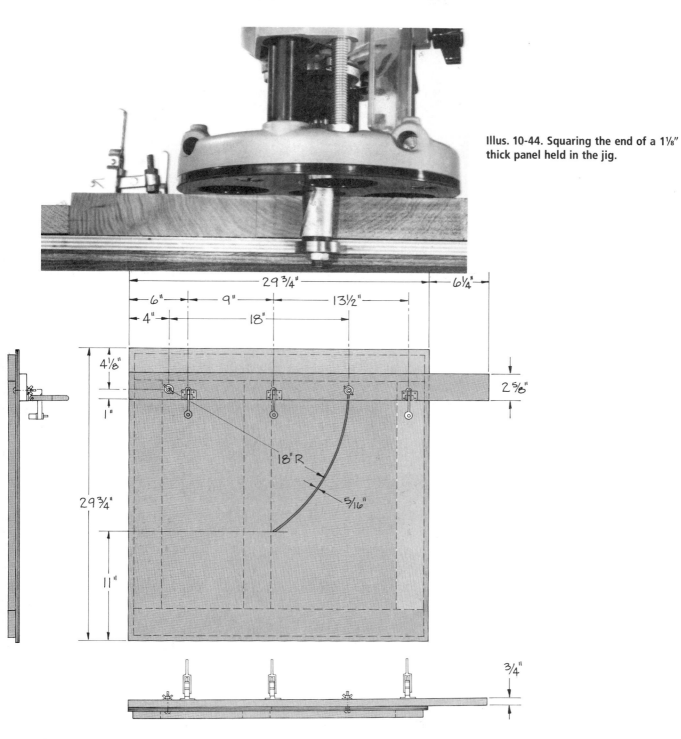

Illus. 10-44. Squaring the end of a 1⅛″
thick panel held in the jig.

Illus. 10-45. Typical details for making a large panel-trimming jig. The hidden lines represent an
optional flat frame cut from ¾″ × 2½″ strips. Its only purpose is to raise the bottom of the base off
the workbench.

11
Routing Joints

Many popular joints and a variety of basic woodworking construction details are presented in this chapter (Illus. 11-1). Virtually any joint can be cut partially or completely with the router (Illus. 11-2). However, sometimes it may not be practical to use the router. For example, you can cut the holes for dowel joints with a plunge router (using only router bits, not drills), but if you own a dowel jig or a drill press, why bother?

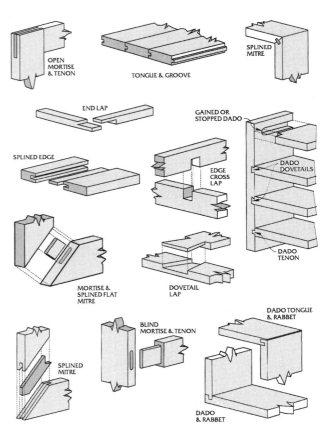

Illus. 11-2. Some joints that can be made with the hand-held router.

Illus. 11-1. This clock project involves lots of joinery and router-cut mouldings.

Depending upon the specific nature of the joint, some joinery may require special bits, specially constructed jigs and fixtures, or a router table to make the job not only possible but safe as well. Later chapters address some of these types of joinery—dovetailing, biscuit joinery, drawer and door joints, raised-panel work, and cope-cut stile-and-rail work.

This chapter covers a variety of basic joints that incorporate tongue, spline, tenon, and mortise cuts. Most are made in combination with the rabbeting, grooving, and dadoing cuts discussed earlier.

The basic joinery cuts, particularly the rabbet, dado, and groove cuts (Illus. 11-3 and 11-4), can usually be cut more quickly and more cleanly with the router than the table saw. Most of the other joints are an adaptation of the basic cuts involved in cutting dadoes and rabbets. A tongue, for example, is simply a rabbet cut made in both surfaces along a common edge. A tenon in the mortise-and-tenon joint (Illus. 11-5) is in very basic terms simply a rabbet cut around the end of one member.

Mortises

Usually, I cut mortises first and then cut the tenons to fit. Mortises can be considered grooves because they are usually cut with the grain (Illus. 11-6). Illus. 11-7 and 11-8 show how easily mortises can be cut using the plunge router with the edge-guide accessory. Mortising with a fixed-base router is more difficult because the rotating bit must be lowered into the work by pivoting or tilting

Illus. 11-5. Not all tenons are intended for great strength.

Illus. 11-3. Dado and rabbet joints are basic cabinet joints.

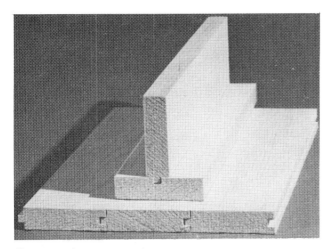

Illus. 11-4. The tongue-and-groove joint.

Illus. 11-6. Shallow dadoes, grooves, and mortises are all about the same width and depth of cut in this side piece for a deacon's bench.

Illus. 11-7. The easiest way to make repeated mortises in stock of the same thickness is with two edge guides, as shown.

Illus. 11-8. Making a pair of stiles with identical mortise cuts. Here the parts are clamped together to provide additional support for the router base.

Illus. 11-9. Routing a center mortise. This cut is best made in two passes with the router guide riding against each side of the workpiece. The resulting cut will then be in the exact center.

the base on the surface. A wide support block attached to the edge guide helps to control the router movement (Illus. 11-9).

A special router base accessory, called the Rig-A-Mortise, automatically centers the cut. This device, available from most woodworking mail-order supply catalogues, replaces the sub-base of plunge and fixed-base routers. The Rig-A-Mortise sub-base has a double guide-pin design (Illus. 11-10). When you feed the router while keeping both pins held against the surfaces of the board (Illus. 11-11), the bit is perfectly centered over the work. Regardless of stock size (up to 6″ in thickness), this device always centers the cut (Illus. 11-12).

The simple shop-made jig shown in Illus. 11-13 is helpful when you are cutting mortises very near the end of the stock. It can also be used to rout mortises into the edges of flat mitres.

Illus. 11-11. How the guide pins center the router and bit over the work.

Illus. 11-10. The Rig-A-Mortise mortising accessory is a plastic sub-base with two adjustable vertical guide pins.

Illus. 11-12. Cutting these perfectly centered mortises on stock of different thicknesses was quick and easy with the Rig-A-Mortise accessory.

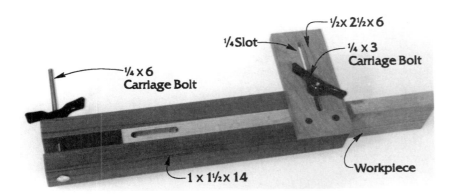

Illus. 11-13. This jig provides extended surfaces for the guide pins. Use it when mortising near the end of a board or into edges of flat mitres.

Remember that slotting bits are great for cutting grooves and slots, and for making various cuts for splines (Illus. 11-14 and 11-15) and biscuit joinery.

Illus. 11-14. A splined mitre joint. On the left are its pieces. On the right is the assembled joint.

Illus. 11-15. Making a splined 45-degree mitre joint using a ⅛″ slotting cutter with ball-bearing guide. Clamp the pieces back to back, as shown. The spline kerf can be stopped, to make a blind-splined mitre.

Lap and Tenon Joints

Lap cuts (Illus. 11-16 and 11-17) can be made fairly easily with a T-square guide. Use a straight bit to make the shoulder cuts to the appropriate depth, as when cutting dadoes. Then clean up between the cuts to get the exact lap width. End lap joints can be used to make quick, inexpensive door and panel frames. Illus. 11-18 shows a simple end lap joint with an edge rabbet suitable for making small doors. Illus. 11-19 shows a good gift project made with scrap material.

Making tenons is similar to cutting laps, but a cut is made into each surface at the end of the workpiece (Illus. 11-20). Fairly short tenons can be cut using the edge guide. Always use the widest bit available and be sure to maintain pressure over the part of the router riding on the workpiece, to prevent the router from tipping over. You want smooth, flat cheeks for the best gluing surfaces. For those who intend to cut lots of tenons, I would

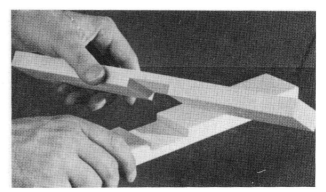

Illus. 11-16. A middle-lap joint.

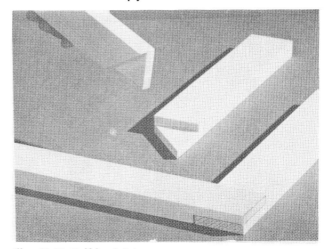

Illus. 11-17. Half-lap joints.

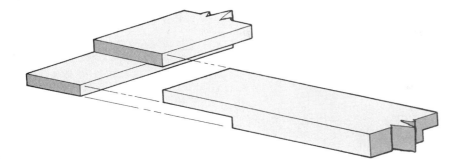

Illus. 11-18. Detail of rabbeted edge frame that has a simple corner lap joint.

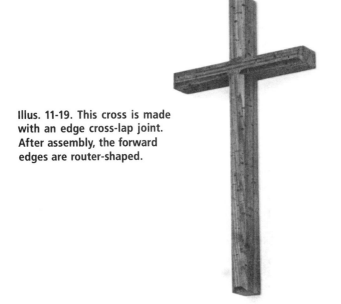

Illus. 11-19. This cross is made with an edge cross-lap joint. After assembly, the forward edges are router-shaped.

Illus. 11-21. Making multiple shoulder cuts with the router guided by a straightedge.

Illus. 11-22. Routing the cheeks of the tenon. Use the widest bit available to give the smoothest and flattest finished surface to the cheeks of the tenons.

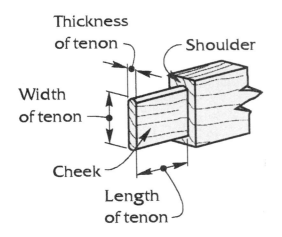

Illus. 11-20. Tenon terminology.

Thickness of tenon

Shoulder

Width of tenon

Cheek

Length of tenon

recommend making a tenon jig similar to the one shown in Illus. 11-23 and 11-24.

Because mortises have rounded, or radiused, ends,

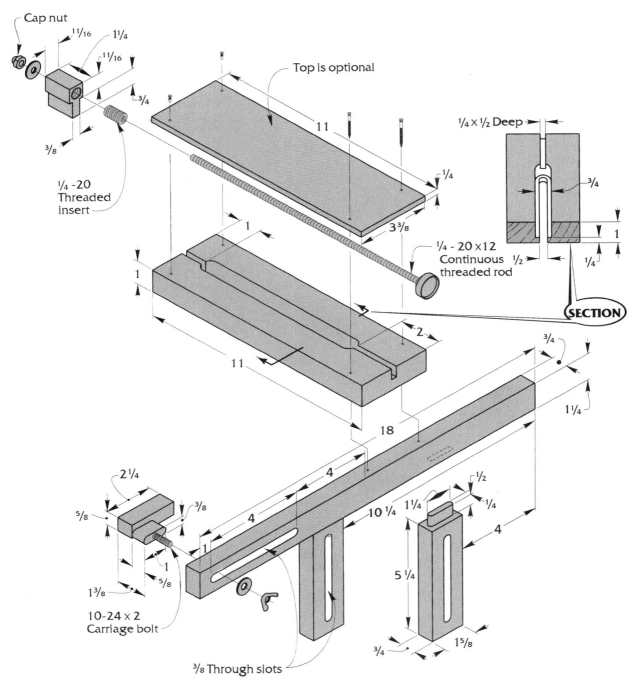

Cap nut

$^{11}/_{16}$ $1^1/_4$

$^{11}/_{16}$

$^3/_4$

$^3/_8$

$^1/_4$ -20 Threaded insert

Top is optional

11

$^1/_4$

$3^3/_8$

$^1/_4$ - 20 ×12 Continuous threaded rod

$^1/_4 × ^1/_2$ Deep

$^3/_4$

$^1/_2$ $^1/_4$

1

SECTION

1

1

11

2

18

$^3/_4$

$1^1/_4$

4

4

4

$10^1/_4$

$1^1/_4$

$^1/_2$

$^1/_4$

4

$5^1/_4$

$^3/_4$ $1^5/_8$

$2^1/_4$

$^5/_8$

$^3/_8$

1

$^5/_8$

$1^3/_8$

1

10-24 × 2 Carriage bolt

$^3/_8$ Through slots

Illus. 11-23. An easy-to-make tenon jig by Ken Hounshell, the inventor of the Rig-A-Mortise router accessory shown in Illus. 11-10 and 11-11. (This illustration and Illus. 11-24 courtesy of Ken Hounshell.)

Illus. 11-24. Ken Hounshell's tenon jig mounted to a workbench.

Illus. 11-25. Rather than squaring the corners of all the mortises to fit the tenons, it's easier and faster to round the tenons, as shown, with a knife, chisel, or file.

resulting from the curved cut of the bit, either the mortise corners have to be hand-chiselled square or the corners of the tenon must be rounded. I prefer the latter technique (Illus. 11-25).

Making Wall Panelling and Wood Floors

A router is a practical and cost-saving tool to use to make wall panelling (Illus. 11-26) and floors (Illus. 11-27). In order to save material, I use edge-spline joints to make wall panelling. A ⅛″ × ⅝″ length of hardboard forms the spline. Use a slotting cutter or a straight bit with a base guide to cut the kerf for the spline if a table saw is not available. The edges were bevelled slightly with a ball

bearing–guided chamfer bit. Construction details for making solid wood panelling are shown in Illus. 11-28. Note that splines can also be used to "end match." This allows for the use of boards of random length and width to produce panelling.

Illus. 11-26. A router is a good tool to use to make wall panels.

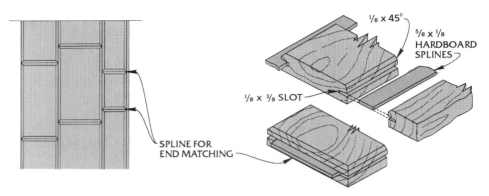

Illus. 11-27 (left). The router can also be used to make tongue-and-grooved floors. Making cuts in the ends (end matching) allows the use of short pieces of random lengths. Illus. 11-28 (above). Details for making splined wall panels.

Cabinetmaking Applications

The router has many applications in cabinetmaking. Furniture can be given that professional look if a router is used with the appropriate bits. Simple mouldings (Illus. 11-29) and those with complex profile sections have many uses for the woodworker. Picture frames, trim around doors and windows, and decorative trim on furniture are all applications of user-made mouldings. It is quite simple to make mouldings. It is best to make narrow mouldings, first shaping the edge of a wider board with the router and then ripping off the moulding using a saw.

Illus. 11-30. The rabbeted back of a picture frame. The same types of cut can be used for making glass or panel cabinet doors.

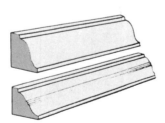

Illus. 11-29. Router-cut mouldings.

Picture frames are made in other ways besides using narrow moulding. Illus. 11-30 and 11-31 show some interesting applications of the router in picture-framing.

Box and Case Work

The serious cabinetmaker puts the router to good use in machining many joints and construction details. Basic box and case construction (Illus. 11-32 and 11-33) involves various rabbet joints. Many times in conventional woodworking practice stopped rabbets must be made

Illus. 11-31. Close-up of a face-inlet, splined mitre joint used in a picture frame. Note the router-shaped edges.

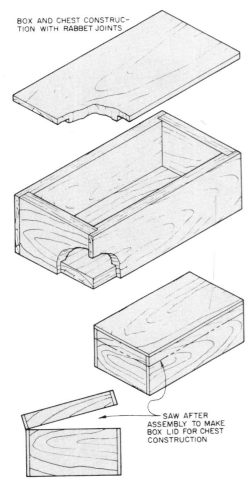

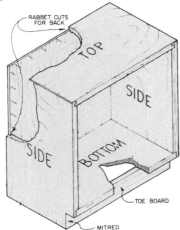

Illus. 11-32. Box and chest construction made with the simple rabbet joint.

Illus. 11-33. Most joints in this cabinet case can be router-cut. Note the rabbet cuts used on the back.

and cut before the assembly of the component parts. I have found that for many jobs I can make the rabbet cuts after assembly. It saves time and energy to cut a rabbet that will receive box bottoms and cabinet backs after assembly (Illus. 11-34). I use a ball-bearing-guided rabbeting bit, which, of course, leaves a rounded inside corner. This can be chiselled square. If the rabbet is left as routed, the corner of the panel must be rounded to fit, as shown in Illus. 11-34.

Illus. 11-34. Routing rabbet cuts after assembly (using a piloted bit) eliminates the setup and machining of stopped rabbets. This technique saves time and is useful when rabbets are needed for cabinet and case backs, box and chest bottoms, and for picture frames. The corner may be left with a radius (as shown here) or chiselled square.

Cabinet-Framing Techniques

Cabinet framing techniques are used to make drawer supports or skeleton framings, cabinet facings, and door frames. Illus. 11-35 and 11-36 show typical joinery without the use of mechanical fasteners.

Pocket Joinery Pocket joinery (Illus. 11-37) is a relatively new and faster cabinet-framing technique that *does* involve screw assembly and eliminates all tenon-and-groove cutting and gluing operations. Pocket joinery has many applications in the workshop. Use it for all right-angled assemblies where only one side (opposite the pocket cut) is visible or needs to be serviceable. Typical jobs for pocket joinery include utility box assembly, attaching furniture rails, aprons, mouldings, corner posts, and base box frames, making panels on frame doors, and even hanging fixed shelves.

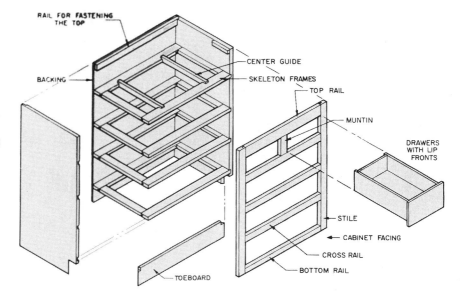

Illus. 11-35. An exploded view of basic cabinet construction.

RAIL FOR FASTENING THE TOP

BACKING

CENTER GUIDE

SKELETON FRAMES

TOP RAIL

MUNTIN

DRAWERS WITH LIP FRONTS

STILE

CABINET FACING

CROSS RAIL

BOTTOM RAIL

TOEBOARD

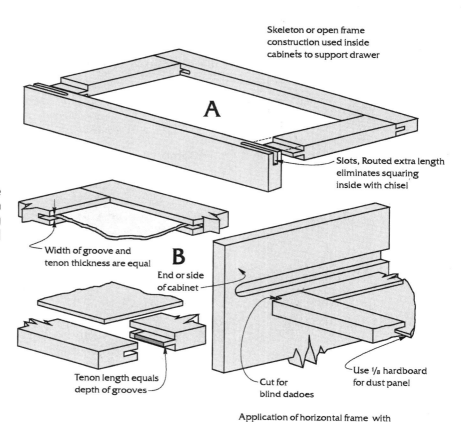

Illus. 11-36. Horizontal cabinet frame members. A: an open frame. B: a framed dust panel (note the machining necessary when stopped [blind] dadoes are used).

Skeleton or open frame construction used inside cabinets to support drawer

A

Slots, Routed extra length eliminates squaring inside with chisel

Width of groove and tenon thickness are equal

B

End or side of cabinet

Tenon length equals depth of grooves

Cut for blind dadoes

Use 1/8 hardboard for dust panel

Application of horizontal frame with dust panel used on cabinet construction

Illus. 11-37. A rail-to-stile pocket joint attached with square-drive panhead screws.

Illus. 11-39. Section drawings of pocket joinery. Above: problems typical of pocket joinery jigs that drill screw holes at an angle. Below: Porter-Cable's routing pocket cutter and the shop-made pocket routing jig are designed so that the screw is installed parallel to the work face.

There are many jig systems you can purchase that involve only drilling operations. Porter-Cable produces a portable routing machine that has some definite advantages over other systems (Illus. 11-38 and 11-39). It makes a router-cut pocket; then you must drill the screw's clearance hole using the clamp and prepositioned drill bushing guide of the machine (Illus. 11-40 and 11-41).

Illus. 11-42–11-44 show a shop-made jig for non-production pocket-routing and drilling that is designed

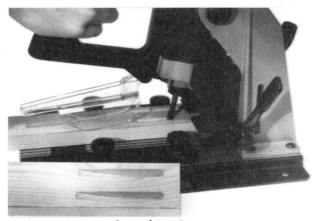

Illus. 11-40. Routing the pocket cuts.

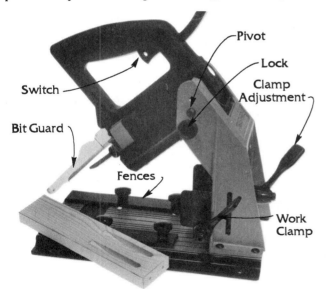

Pivot

Lock

Clamp Adjustment

Switch

Bit Guard

Fences

Work Clamp

Illus. 11-38. Porter-Cable's pocket cutter is powered by a 5-amp router motor with a pivoting action. A 5⁄16″ diameter bit sweeps in a long arc, cutting into the work. The unit also incorporates a system of stops, clamps and a built-in guide bushing for drilling the screw clearance hole.

Illus. 11-41. Drilling the 7⁄64″ screw clearance holes.

Illus. 11-42. Router and jig design for pocket-cutting. Note the slot and the two viewing holes cut in the jig. The router base carries a template guide and a ⅛″ thick piece of wood fastened with double-faced tape so that the router can ride on the curved, inclined surface of the jig.

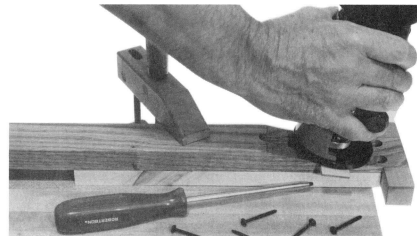

Illus. 11-43. Making the pocket cut with the jig clamped to the stock.

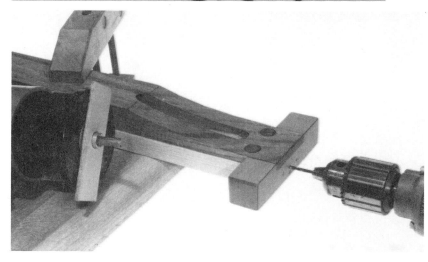

Illus. 11-44. Use a ⁷⁄₆₄″ drill guided by the jig to make the screw clearance hole(s) in the workpiece.

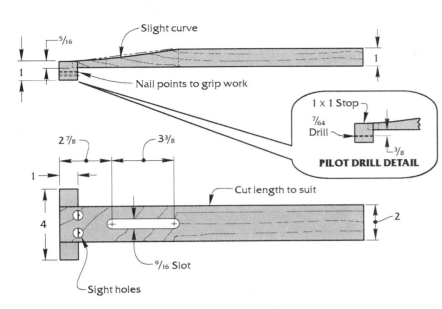

Illus. 11-45. Pocket-routing jig details. This one is designed for a Porter-Cable ½" outside diameter template guide and a ¼" diameter router bit.

for use with a trimming or light-duty router. The general construction details are given in Illus. 11-45. You may need to modify the plan somewhat to accommodate your brand and size of template guide (see Illus. 11-46).

Drawers and Doors

Rabbet cuts and grooving operations are also necessary for drawer and door construction in furniture-making. Illus. 11-47 gives basic specifications and assembly details for typical lip-and-flush drawers.

Cabinet doors are made in many different styles and with many various decorative touches. I have included some basic types, to illustrate fundamental construction

details and design suggestions (Illus. 11-49 and 11-50). Looking through cabinet and furniture catalogues will give you other design ideas. Most can be fabricated by employing the joints and assembly methods given in this chapter. Refer to Chapter 23 for details about making raised-panel doors with coped stile-and-rail frames, and Chapter 18 for more door design ideas involving panel-routing techniques.

Edge-Banding

In much of today's fine furniture-making, plywood or other sheet material covered with beautiful veneers is used (Illus. 11-51). Some ways of covering the edges of veneer-faced plywood and other sheet material are shown in 11-52. Most applications are some form of a tongue or groove cut into a solid edging. Once a solid edge is applied, almost any shape can then be routed to the edge.

Drop-Leaf-Table Joint

The drop-leaf table joint (Illus. 11-53) is commonly used to make a hinged table leaf. A cove bit is used to make the cut in the leaf. A rounding-over bit of the same radius set to the proper depth makes the cut in the tabletop. A small core box bit is used to mortise the underside of the tabletop for the hinge barrels. Do not attempt to mortise the hinge itself into the work.

It is best to make practice cuts on scrap stock of the same thickness as the finished table, and to mount hinges to scrap pieces, and to check all cuts. It is essential that the distance from A to B in Illus. 11-53 be exactly the

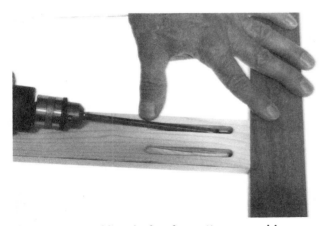

Illus. 11-46. Assembling the face frame. Use square-drive pan-head screws or Phillips® screws 1¼" × 1½" in length.

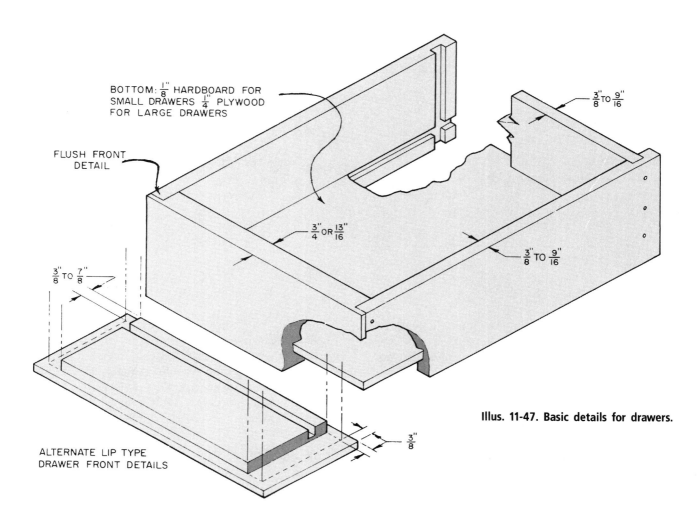

BOTTOM: $\frac{1}{8}$" HARDBOARD FOR
SMALL DRAWERS $\frac{1}{4}$" PLYWOOD
FOR LARGE DRAWERS

FLUSH FRONT
DETAIL

$\frac{3}{8}$" TO $\frac{9}{16}$"

$\frac{3}{4}$" OR $\frac{13}{16}$"

$\frac{3}{8}$" TO $\frac{9}{16}$"

$\frac{3}{8}$" TO $\frac{7}{8}$"

$\frac{3}{8}$"

ALTERNATE LIP TYPE
DRAWER FRONT DETAILS

Illus. 11-47. Basic details for drawers.

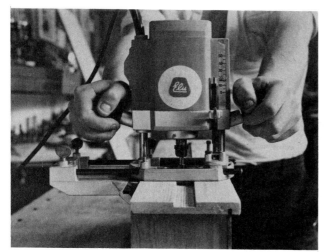

Illus. 11-48. Grooving a drawer side with an edge guide.

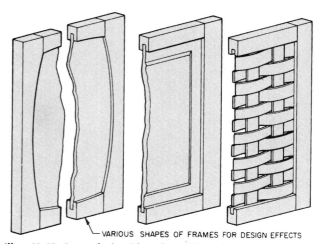

VARIOUS SHAPES OF FRAMES FOR DESIGN EFFECTS

Illus. 11-49. Some design ideas for cabinet doors.

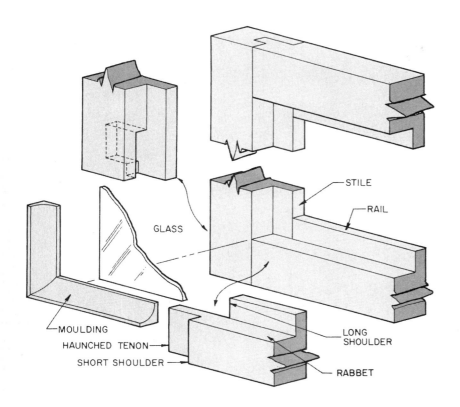

GLASS

STILE

RAIL

MOULDING

HAUNCHED TENON

SHORT SHOULDER

LONG SHOULDER

RABBET

Illus. 11-50. Details for making medium-to-large glass- and/or panel-inset doors with typical mortise-and-tenon construction.

Illus. 11-51. This handsome teak dining table by Scan Furniture features wood-sheltered hinges and veneer-faced material with solid wood edging and legs.

Illus. 11-52. Some ways of covering the edges of plywood or veneered particleboard.

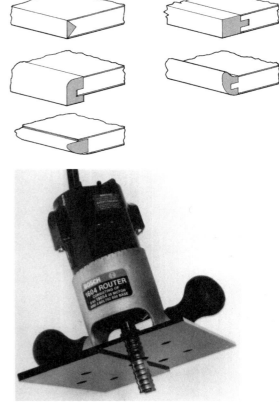

Illus. 11-53. A drop-leaf table joint. The distance from A to B must equal the distance from B to C. Note that only the barrel of the hinge is mortised, not the leaves.

same as that from B to C. To make the full edge-shaped cut, use the piloted rounding-over bit with a straight-edged board clamped under the work for the pilot to ride against. Set the core box bit about 1/32" deeper than required, to obtain clearance between the leaf and the top.

Solid-Panel Joinery

Large panels made of narrower pieces glued together at the edges are found in some of the finest furniture. Several router-bit companies provide special cutters and stepped router bases (Illus. 11-54) to make matched interlocking joints for edge-to-edge glue-ups of wide panels such as those used on solid-wood tabletops. These "systems" were primarily developed for use in fabricating solid surface materials (Corian, Avonite, etc.), but they are also great for solid-wood glued edges. They make a wave, or shallow finger-joint, cut that helps to align the joint and give it more gluing surface, for a stronger joint (Illus. 11-55 and 11-56). The male portion of the cut produced on one edge matches the female or inward portion of the cut on the edge of the mating piece.

Illus. 11-54. The Bosch edge-jointing system includes a special patterned bit and a stepped router base. The stepped edge of the base is *not* used as a guide for the router.

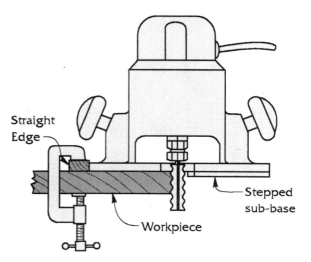

Illus. 11-55. Porter-Cable's Tru Match edge-joining system. Note that the edge of the router base follows a clamped straightedge while making the cut.

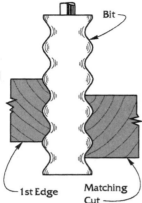

Illus. 11-56. The bit is referenced to the surface of the base so one element of the patterned bit is divided or bisected.

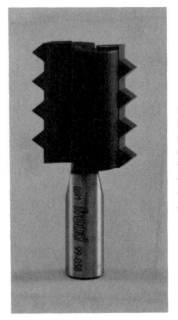

Illus. 11-57. Freud's V panel bit. In use, the bit does not need to be centered on the edge when you are routing the first edge. The bit is then raised or lowered $\frac{7}{32}''$ to rout the mating edge on the second piece.

Once the bit's pattern-to-base depth of cut has been set correctly, make two passes, one along each edge of the two pieces of the joint, without changing the bit's depth of cut. The cut on the mating piece is made with the other stepped surface of the router base riding on the workpiece surface. The edge of the router base must follow against a straightedge clamped to the workpiece for each cut. Use the opposite edge of the stepped base when making the cut along the second or mating piece.

One advantage of using these edge-joining systems is that when the pieces are assembled you will have a flush, flat working surface even if the two mated pieces are of different thicknesses.

Freud sells a V panel bit (Illus. 11-57 and 11-58) that, like other finger-joint cutters, requires that you change its cutting depth to make the matching cut on the mating piece.

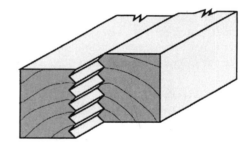

Illus. 11-58. A joint made with Freud's V panel bit.

12
Surface-Routing Techniques

Essentially, this chapter examines a variety of techniques involved in two fundamental routing applications: surfacing (flattening or levelling a surface) and making decorative straight-line cuts into flat surfaces. Depending upon your own specific job requirements, the hand-held router may be guided by a variety of devices such as straightedges, edge guides, or specially made jigs and fixtures. In this chapter a number of unusual jigs are illustrated which may inspire you to make one for your own particular requirements. The next chapter covers template- and pattern-routing techniques for making and duplicating irregular surface cuts.

Preparing Stock to a Parallel Thickness

Surfacing stock to a parallel thickness can only be done on a very limited basis with a router. However, in some situations the router is more practical and functional than a surface planer. One such use is to surface old,

varnished, or painted lumber that you don't want to run through your planer because it can easily dull high-speed-steel knives.

The fixture shown in Illus. 12-1 will work perfectly when used with a suitable carbide bit (Illus. 12-2). You can make your jig to accommodate special needs. I seldom use solid stock wider than 6″ to 8″ in width, so the one shown in Illus. 12-1 satisfies my needs. The fixture has two level router guide rails; one is horizontally adjustable with vertical dowel pins. The work is held by two wedges that force the workpiece edge against pointed nails so that it doesn't slip during routing. The router is mounted to a long, extended base made of rigid plywood. Stiffeners can be added on top of it, if necessary. Illus. 12-3 contains construction details for making the work-holding fixture.

Rough-cut log slabs can also be levelled with the same fixture, but a more quickly made box or frame (Illus.

Illus. 12-1. This adjustable surfacing fixture can be made to handle stock of almost any width or length.

Illus. 12-2. Any wide, flat-bottom cutting bit can be used to make surfacing cuts. The wider the bit, the better. At left are ¾″ and 1½″ diameter dado planer bits. At right is a 2¼″ diameter drawer lock bit, which should not turn faster than 16,000 rpm.

12-4–12-7) will also keep the router travelling level over the log slabs. Secure the slab with hot-melt glue, nails, or wedges, or any other way that works.

Routing Flat, Tapered Surfaces

Flat, tapered surfaces such as those that might be required to make square, tapered table legs can be routed in a couple of ways. For wider pieces with slight tapers, you can modify the jig shown in Illus. 12-1 and 12-3 by inserting a block or wedge under the workpiece. When routing square or otherwise narrow tapers, you can make a special work-supporting fixture as shown in Illus. 12-8 and 12-9. To repeat the same cut on different pieces or surfaces of the same piece, mark and reference the small end of the workpiece to the inside lip of the jig or to

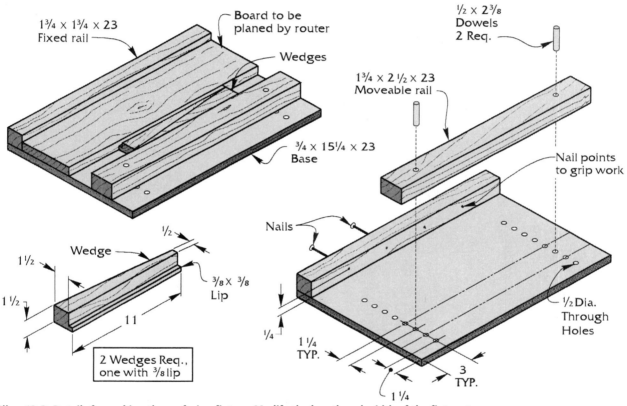

Illus. 12-3. Details for making the surfacing fixture. Modify the length and width of the fixture to suit your needs.

Illus. 12-4. Some boards used as extended router bases. The top two have integral template guides; one is made of aluminum tubing, and the other (center) of plastic pipe. The lowest one is used for general surfacing, as shown in Illus. 12-1 and 12-5.

Illus. 12-5. Levelling a thick chunk of wood in preparation for bowl-turning in the lathe. Here a simple box frame keeps the router (with an extended base) level.

Illus. 12-6. Levelling a log end with a router.

Illus. 12-7. A close-up of a log end being levelled.

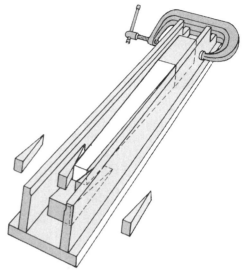

Illus. 12-8. A fixture for cutting the slanted surfaces to make square, tapered furniture legs.

Illus. 12-9. The same fixture may be used to hold the tapered leg for fluting (shown) or other surface-decorating jobs.

another mark. A reference point system is also necessary if the workpiece will be reinserted in the fixture so that you can make subsequent decorative flute or grooving cuts in the taper-cut surfaces.

Decorative Grooving, Beading, or Fluting

Surface cuts made parallel to each other create an interesting textured effect. Using point-cutting ogee and quarter-round bits (both available from Sears) and any straight-line router guiding system will produce the results shown in Illus. 12-10. Other results are achieved using cove-cutting, V-groove, or small veining bits (Illus. 12-11 and 12-12).

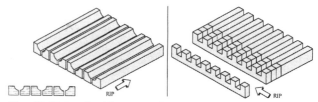

Illus. 12-11. Stock with accurately spaced, multiple surface cuts can be sawn into narrow mouldings.

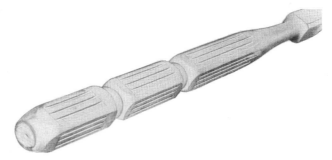

Illus. 12-12. An example of a turned leg with parallel-fluted grooves on the flats.

Illus. 12-13 and 12-14 show a production technique I used to run decorative center-grooves in hundreds of balusters made for railings around wood decks. The jig incorporates three stop blocks. They not only allowed me to rout the pieces without clamping the workpiece, but

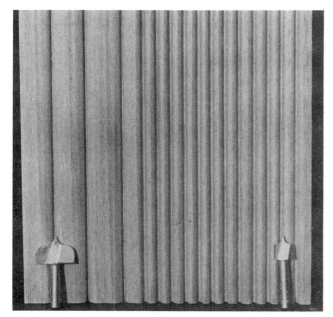

Illus. 12-10. Parallel cuts in surfaces create a textured effect.

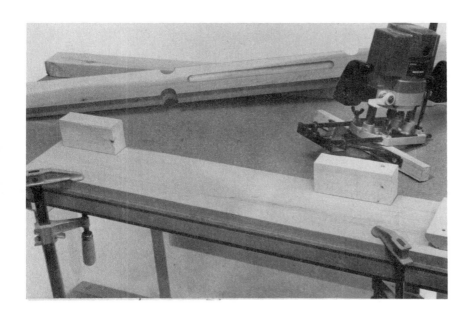

Illus. 12-13. This jig is used for the production-routing of single, cove-shaped stopped grooves in deck balusters (shown in the background).

two blocks provided the start and stop locations for the plunge-routed, cove-cut grooves (Illus. 12-15). The workpiece does not move or shift under the router because the side and forward pressure applied keeps it in place against the stops.

Right-Angle and Oblique Cuts

Right-angle and oblique cuts can be all made into the surface from one side or cut into both surfaces to create a wood grille effect (Illus. 12-16). Some intriguing-looking projects can be developed if you employ these interesting routing techniques (Illus. 12-17 and 12-18).

By varying the depth of cut, the direction or angle of the cuts, and the cutting profile of the bit used to make the cut, you can create a multitude of designs (Illus. 12-19).

Radial Jig

A radial jig (Illus. 12-20 and 12-21) is a versatile, easy-to-make accessory. This device consists simply of a parallel-

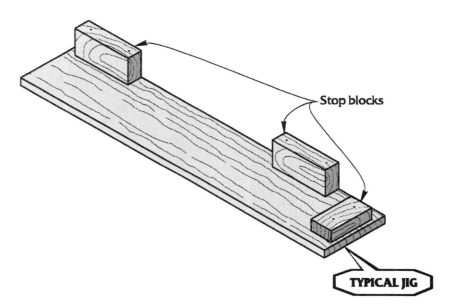

Illus. 12-14. The fixture consists of three stop blocks and a base.

Illus. 12-15. The two side blocks start and stop the grooving cut.

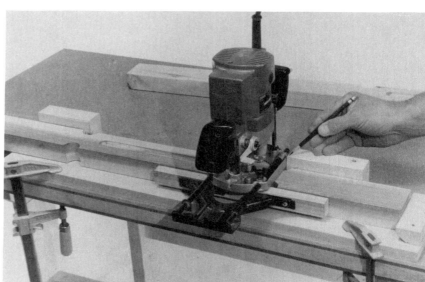

tracked base or guided router slide that is slotted for the bit. It is raised above the workpiece and pivoted on one end with a single flathead wooden screw driven into a work fence. The particleboard base of the jig shown in Illus. 12-20 and 12-21 is 4′ long and 35″ wide. It has a radius cut along one side to facilitate clamping the 35″ long router slide at any angle to the 2″ thick work fence.

This shop-made routing guide can be used in many job applications. It can be used to cut grooves in panels, to cut dadoes, and for decorative routing such <u>as that</u> done to make the slotted grilles shown in this chapter. The possibilities can be further increased if you use the jig with a plunging router and clamp stop blocks to the router slide to limit horizontal router travel. With this setup, it is possible to make stopped dadoes, stopped grooves, etc.

Illus. 12-16. Right-angled grooves cut into both surfaces create this interesting grille pattern. The cutting depth is <u>greater</u> than half the stock thickness.

Illus. 12-17. Grilles of router-cut boards are assembled to make this lamp.

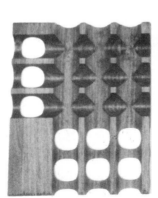

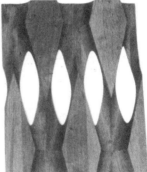

Illus. 12-18 (above left). These trivets are created by routing from both sides to half the stock thickness. Illus. 12-19 (above right). More examples of decorative grille work possible with the router.

Illus. 12-20. A radial router accessory. Note the spacing block under the clamp and the board supporting the router slide. This device will handle work 2″ thick. Thinner stock can be raised with shim stock to reduce the reach of the router.

Illus. 12-21. When the radial router jig is used with a secondary 90-degree fence, it has adequate support for many angle-routing jobs.

A plunging router is also ideal for use with this jig where deep multiple or identical cuts are necessary. Most plunging routers can be set with multiple depth stops. Thus, when several passes are required to make a deep cut, the final pass will always result in exactly the same depth as the preceding cuts.

One problem that you must overcome when using this jig is the workpiece's tendency to slip during routing. Various clamping methods have to be devised, especially when you are routing short pieces. Small workpieces can be tacked and then glued with hot-melt glue or nailed to larger and longer support boards that, in turn, are clamped to one side of the base. Pointed nails extending through the fence will hold the stock, too, but they make

holes in the edge of the workpiece. A secondary fence (Illus. 12-22) is often helpful for many jobs, especially cuts that are sharply angled to the edge of the workpiece.

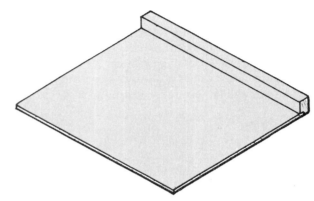

Illus. 12-22. A secondary fence for the radial router jig. It is clamped to the base to make another work fence that is 90 degrees to the primary fence.

Serpentine Routing

Illus. 12-23 shows a contoured surface being routed. The simple fixture in the photograph is used to make grooves or flutes that wind in one direction and then another. Hence, this kind of work is called serpentine routing. The construction details of the fixture are fairly straight-forward. A wood V-block or bandsawn sled-shaped cradle holds the router motor in a horizontal position. Cut a large hose clamp in half and screw each end to the motor-mounting block.

Illus. 12-23. Serpentine routing with a simple device supporting the router horizontally. Note the guide-pin follower that ensures uniform depth of cut.

The guide pin or follower is simply a ¼″ diameter machine bolt with its head cut off that has been bolted through a counterbored hole in the bottom of the base. Note that in Illus. 12-23 the workpiece is raised with some scrap boards so that the cut is made at the desired location. Some stock bits with ball-bearing guides, like the one shown in Illus. 12-24, can be used without any jig or fixture to make decorative bead, cove, and grooving cuts at a uniform depth.

Illus. 12-24. Ball-bearing-guided bits can be used to make grooving and beading cuts at uniform depths.

Illus. 12-25 takes serpentine routing a step further, combining horizontal routing with a template guide. The template can be of any configuration. This principle may be applied to machining decorative grooves into flat or gradually contoured surfaces. In both instances, you can place pieces of plywood under either the workpiece or the

router-supporting fixture to position the work in relation to the bit so that the cut can be made exactly where desired. To make the cut, clamp the work in position and slide the router fixture along the surface of the workbench. More sophisticated fixtures and operating techniques can be used to satisfy individual needs. For production work, a ball bearing can be mounted to the guide pin or attached to the pointed follower. A combination workpiece-positioning fixture with stops, integral clamps, and a template or pattern can be added as desired.

Illus. 12-25. Template-guided routing produces tapered grooves in a flat surface. Note the template and pointed follower at the base of the fixture, and the shims between the template and the work.

Illus. 12-26. The basic necessities for router-carving wheat patterns include a trimming router with a base-mounted template guide, a roundnose bit which cuts as shown, and an inclined jig block.

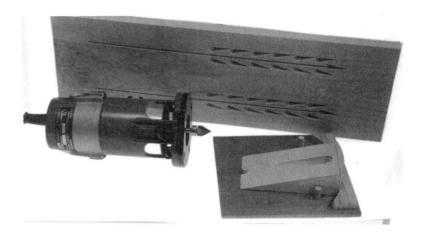

Illus. 12-27. A ball-bearing-guided V-bit cuts a slightly different wheat design. Note the pivoting inclined guide block with the two dowel stops.

Routing Wheat Patterns

I've developed two fairly easy methods of carving decorative wheat designs into flat panels. Since these are shallow cuts, a trimming router that you can operate one-handed works best. The problem is to guide the router so the cuts can be consistently made. You can control the direction of the router either by mounting a tubular template guide into the router base through which the bit protrudes (Illus. 12-26) or by using a ball-bearing-guided bit (Illus. 12-27). With both methods, a slot in the wedge-shaped block and the router depth-of-cut adjustment control the length of each teardrop-shaped cut which comprises the wheat pattern. When used alone, the wedge block can be repeatedly positioned right and left with the aid of a set of marks located on its edges. These marks correspond to incremental spaces marked

off along a centerline on the pattern design, thus maintaining a consistent angle for each cut and the spacing between them (Illus. 12-28). Two strips of sandpaper (60–80 grit) applied (with double-faced tape) to the bottom of the slotted, wedge-shaped guide block keep it from slipping during the cut (Illus. 12-29).

A slightly more advanced version of this block that works well on longer designs includes a base with a pivot point and right-and-left-cut stops. To use this jig, hold it, make one cut on one side of the centerline, and, without moving the jig, pivot the inclined block to the other side to make the mating cut. If you are cutting large, flat surfaces, apply abrasive paper to the bottom of the base to help hold the jig. When cutting smaller panels, however, do not apply abrasive paper to the inclined guide block (Illus. 12-30). You can adapt this jig so that you can make the wheat design follow curved stems or lines.

Illus. 12-28. Marks on the edge of the jig and its opposite lower corner are aligned with a pattern centerline marked on the work. This line is stepped off in distances that equal the space between cuts.

Illus. 12-29. Strips of sandpaper applied with double-faced tape keep the guide block from slipping under the router.

Illus. 12-30. When the inclined block is used with a base for a pivot and stops, apply abrasive paper to the base, as shown, not to the inclined guide block.

Illus. 12-31. Here you see the jig with two optional outriggers, or saddles, that straddle this particular workpiece. Spacing for each pair of cuts is stepped off along the edge to ensure uniformity of the pattern.

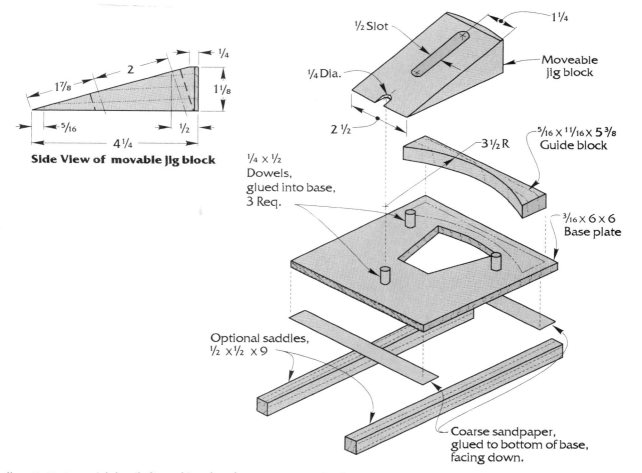

Side View of movable jig block

Illus. 12-32. Essential details for making the wheat-pattern routing jigs.

Angle Routing

I've always wanted a plunge router with a base you can adjust or tilt to any angle, like those found on a portable circular saw. Something that is almost as good is the adjustable cast-aluminum clamp bracket for router motor units made by Bryco, Inc. of Champaign, Illinois (Illus. 12-33). This device will prove extremely helpful to those who want to make unusual straight-line textured surfaces (Illus. 12-34 and 12-35). Heretofore, routers only approached the work's surface with vertically orientated bits. Now, when you use this tilting mechanism, one bit can produce a variety of cutting configurations because it's possible to approach the work from a variety of different positions or angles.

Illus. 12-33 (above left). Bryco's adjustable aluminum router bracket can hold your router's motor unit while it is tilted as much as 45 degrees to the work. Illus. 12-34 (above right). The Bryco router bracket mounted to a simple, square plywood base. Note the shape of the cut(s) produced with a bowl or tray bit when it is angled to the work's surface.

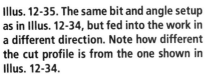

Illus. 12-35. The same bit and angle setup as in Illus. 12-34, but fed into the work in a different direction. Note how different the cut profile is from the one shown in Illus. 12-34.

13
Template- (or Pattern-) Routing

Shop-made templates, or patterns, as they are also called, are used to: (1) guide the router to make irregular (Illus. 13-1 and 13-2) or straight-line cuts into a workpiece (Illus. 13-3) or completely through it; (2) rout or trim edges (Illus. 13-4 and 13-5) of rough-sawn profile shapes; (3) do special joint work such as some kinds of dovetailing, finger joints, etc; (4) set in hardware such as hinges; and (5) assist in making some inlays.

Since dovetailing, setting in hardware, and inlays are covered in other chapters, just the basic functions of template- or pattern-routing are explored here. Many of the essential principles involved in template- and pattern-guided routing with a portable hand-held router

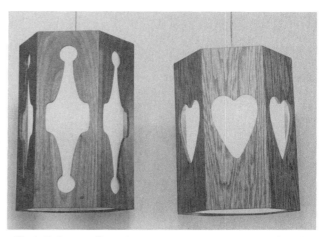

Illus. 13-2. Template- (or pattern-) routing was employed to pierce-cut the identical segments for these hexagonal hanging box lamps.

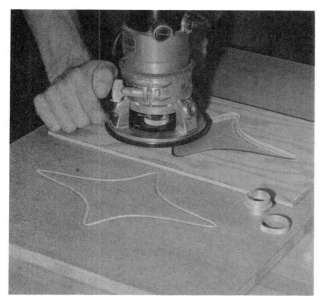

Illus. 13-1. Template-routing makes it easy to precisely duplicate router cuts of any desired design or pattern. Here a router template guide (lower right) attached to the base guides the router.

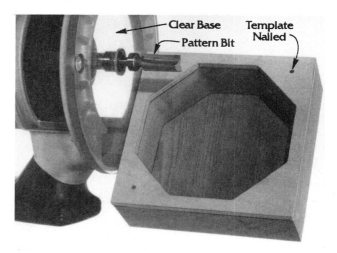

Illus. 13-3. Routing deeply into an octagonal trinket box using a ball-bearing-guided panel bit. Most of the waste was bored out with a Forstner bit before routing.

Illus. 13-4. Applying double-faced tape to a quarter template made of plastic that was used to produce the country craft wheel shown on the top left of the photo. Note the rough-sawn blank with the inside of the hearts bored out to remove excess waste.

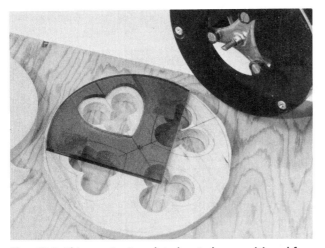

Illus. 13-5. This quarter template has to be repositioned four times to trim the inside and outside edges of each wheel. A small hole in the template fits over a short brad driven into the middle of the wheel, and four reference lines align the middle of the hearts. Note the template guide in the Sears router.

can also be applied to template work done on a router table. Therefore, refer to Chapter 23, which describes router table work, for more information that may be helpful for hand-held work.

Pattern- and template-guided routing used in a hand-held or router-table application is one of the most exciting and satisfying routing techniques. This class of work

affords you the opportunity to use and expand your creative talents as you discover new ways to use patterns and templates.

Though template- or pattern-routing will also prove helpful if you need only to make that one tricky part for a specialized project, its major advantage is that you can easily reproduce templates and patterns on a production basis. Once the design is developed and the template made, you can duplicate precisely cut after cut and project after project. And, this technique isn't complicated or difficult to learn.

To avoid confusion, the terms pattern and template will be used interchangeably. Technically, a *pattern* is an original part or object from which another part or object is precisely traced or copied in its exact shape, size, and detail. A *template* is a guide (or pattern) which is used to duplicate and/or guide routing cuts (Illus. 13-6). Templates are made of various materials, such as plywood, hardboard, solid lumber (Illus. 13-7), plastic, and aluminum. Generally, the type of template material used is determined by convenience, availability, and the desired length of service required. Templates can vary in thickness from ⅛″ to ¾″, depending upon the type of guide attachment fitted to the router base, the depth of the cut required, and the type of bit and length selected for the job. Generally, templates that are ¼″ thick are suitable for the majority of jobs using base-mounted template guides. It is a good idea to make thinner templates when possible, because they are easier to cut and to smooth the edges of (Illus. 13-8 and 13-9). Templates may be cut freehand with the router if a scroll saw or band saw is not available.

There are various kinds of commercially made templates available. These include lettering templates for

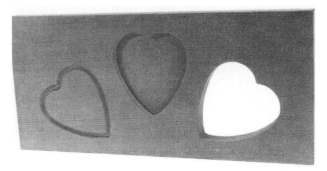

Illus. 13-6. The same template was used to make these three different kinds of cut.

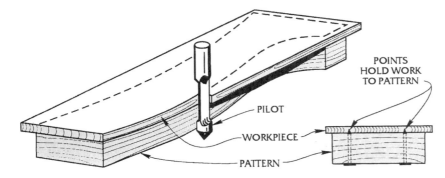

Illus. 13-7. Routing a solid-wood pattern using a piloted panel bit.

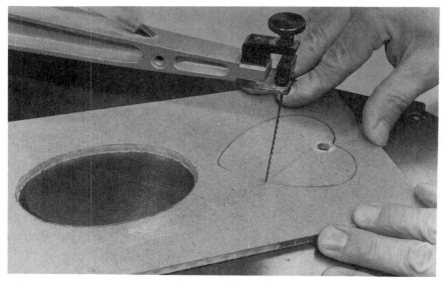

Illus. 13-8. Cutting a ¼″ thick tempered hardboard template with the scroll saw. Note the sharp, accurate layout lines.

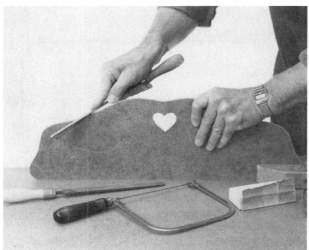

Illus. 13-9. Make the edges smooth with flowing lines because any error or miscut along the edge will be automatically transferred from the template directly to the workpiece.

Illus. 13-10. The Pinske large squaring (90 degree) template is made of heavy aluminum. Similar templates made in 22½ and 45 degrees are available from the Pinske Edge Co., in Plato, Minnesota.

sign work, templates for recessing butt hinges in doors, and even large aluminum templates for special jobs such as squaring and radiusing large panels (Illus. 13-10 and 13-11). To do pattern routing on panels as shown in Illus. 13-7, use piloted or ball-bearing-guided pattern-cutting or trimming bits. When using these bits, prepare the patterns/templates with their profile shapes made to exactly the same size as is desired for the copied part.

Illus. 13-11. Routing an inside 1½″ radius to trim countertop material using a heavy-duty, solid-aluminum template developed and manufactured by Tom Pinske.

Be sure to consider the restraints that thin template material might impose when certain kinds of bits are used. With some pattern-cutting bits, their cutting edges may be too long or just too far from the pilot, so that making multiple passes at shallower depths of cuts is impossible (Illus. 13-12). One problem cutting tool is the popular dishing or tray bit fitted with a shank-mounted bearing. Illus. 13-13–13-15 show the bit set up and used for template routing with fixed and plunge routers.

Illus. 13-12. With this bit design and thin pattern material, making successive passes to arrive at a desired depth is impossible. In this instance, you will need to make your pattern from thicker material.

When using pattern-cutting bits (Illus. 13-16) that have cutting edges too long to permit shallower cuts into the surface of a workpiece, simply insert appropriate shim material under the template to raise it off the work's surface.

When cutting or trimming outside edges of irregular workpieces, it's best to saw the workpiece slightly (⅛″ to ³⁄₁₆″) oversize (Illus. 13-17). Depending upon the intended use of the workpiece or whether it will be visible in the finished product, the template can be taped or nailed to the work blank.

Illus. 13-13. The dishing/tray cutting bit with a shank-mounted bearing must make its initial cut very deep if a thin template is being used. When set up as shown with a thick extended plastic sub-base, it is best used for finished cutting after most of the waste has been hogged out by other wood removal techniques.

Illus. 13-14. Preparations for dishing out a tree tray project with a more powerful plunge router. Note the thick template/pattern secured to the work with double-sided tape.

Illus. 13-15. Routing the inside of the tree tray. The ball bearing on the bit limits the horizontal depth of cut. Note that the work is done while on a router mat, eliminating the need for conventional clamps that would obstruct router movement.

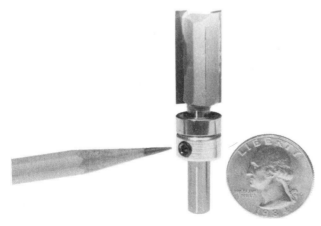

Illus. 13-16. A regular straight bit with a shank-mounted bearing that matches the cutting diameter of the bit makes a good pattern-cutting bit.

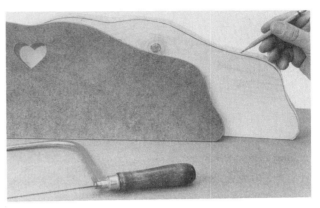

Illus. 13-17. When cutting or trimming outside edges of irregular workpieces, it's best to first cut the workpiece slightly oversize to a rough-sawn shape.

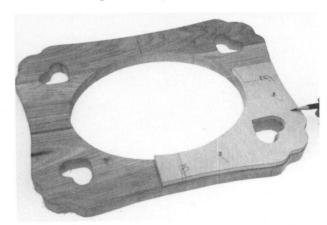

Illus. 13-18. Here, a quarter template for making a solid-wood picture frame is tacked to the back surface, which will not be visible.

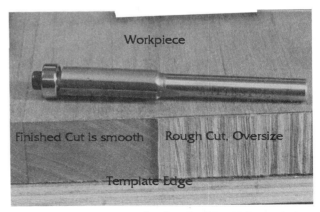

Illus. 13-19. The ever versatile trimming bit cuts the workpiece to the same size and contour as the template.

Template Guides

Template guides are tube-shaped attachments fitted to the center of the router's sub-base (Illus. 13-20). The bit extends through and clears the inside of the hollow tube (Illus. 13-21 and 13-22). The outside walls of the template guide bear against the template or pattern and control the horizontal movement of the router. Template guides are specified according to their inside and outside diameters.

Illus. 13-20–13-26 show a few of the template guides sold by various manufacturers. The style and sizes of

Illus. 13-20. This two-part template guide, similar to that found on Porter-Cable routers, is available in many sizes that range from ¼" to 1¹⁄₃₂" in inside diameter. The guide attaches to the router sub-base with a common lock nut (top illustration).

Illus. 13-22. The bit must clear the inside of the template guide. Here, a dishing bit is checked for clearance through one of the three template guides available for Sears routers. The template guides are available in inside diameters of ⁵⁄₁₆", ⁷⁄₁₆", and ⁵⁄₈".

template guides provided by Porter-Cable have more or less set the standard. Many other router manufacturers, including most foreign ones, are providing special adapters for their sub-bases so that Porter-Cable template guides can now be used with them. Bosch provides the quickest-changing template-guide system of those available (Illus. 13-27 and 13-28). It would be much less confusing if the router manufacturers standardized sizes and styles of template guides, but unfortunately this will not happen.

There is a broad range of template-guide sizes available, but it's my opinion that there still are not enough sizes to select from. It would be better if the center holes of router sub-bases were larger, and larger-diameter template guides were available to accommodate bits with larger cutting diameters. I have one or two of every size template guide available, but I still find myself boring the inside hole slightly larger or cutting the template guide to reduce the length protruding below the router base.

Illus. 13-21. The bit extends through the tubular template guide.

Illus. 13-23. A one-piece template guide such as is used in Black & Decker routers and other routers. Available in many sizes, this type is also fastened to the sub-base with screws.

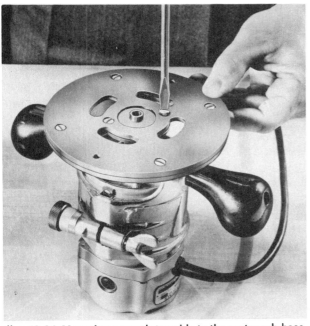

Illus. 13-24. Mounting a template guide to the router sub-base.

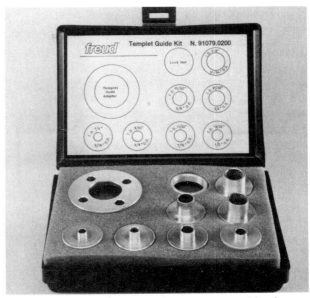

Illus. 13-25. Freud's template guide set comes with adapters that convert the template guide to seven different inside diameters that range from ¼" to 2¹⁄₃₂".

Illus. 13-26. An entirely different template guide system for the new Bosch routers that is designed to be installed and removed quickly. The template guides have an inside diameter that ranges from ⁵⁄₁₆" to 1⁷⁄₆₄".

Illus. 13-27. Spring-loaded locking tabs built inside the base on the Bosch routers engage the template guide.

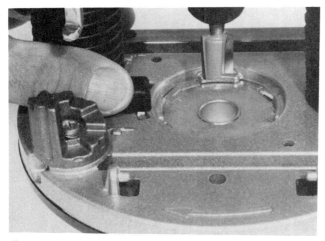

Illus. 13-28. The spring-loaded lever facilitates quick template-guide changes on Bosch routers.

There is one minor problem associated with the use of templates and template guides. Because the bit is encircled at the router base by the template guide, the template or pattern size must be adjusted to allow for the distance between the cutting edge of the bit and the outside diameter of the template guide. This allowance is shown in Illus. 13-29 and 13-30.

Obviously, the bit must clear the inside of the template guide. So, in planning a job, use a bit with a correct diameter with a template guide that has a suitable inside diameter. Remember, too, that the larger the outside diameter of the template guide, the less intricate the detail it will be able to reproduce.

Another thing to consider is the relationship between the thickness of the template material and the distance that the template guide extends below the router base. Consider these two dimensions before making any cuts.

It's usually easier to start a cut holding the template guide against the template edge without the router bit already in contact with the work. Here is where plunging routers have a great advantage. They can be held so that the template guide is bearing against the edge of the template before the bit is lowered into the work. When using a fixed-base router, you must tip it on its base as shown in Illus. 13-31 and carefully lower the rotating bit into the work as close as possible to the desired location without accidentally cutting into the working edge of the template/pattern (Illus. 13-32).

Illus. 13-29. The allowance between the template guide and the cutting edge of the bit. This distance must be subtracted all along the contoured edges of the template.

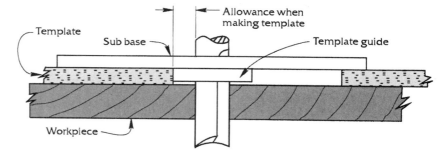

Illus. 13-30. You can measure or calculate the guide to bit allowance. This Porter-Cable template guide has a $\frac{7}{16}''$ outside diameter. When using this template guide with a $\frac{1}{4}''$ diameter bit, allow $\frac{3}{32}''$ when designing a template for a specific-size cut or opening.

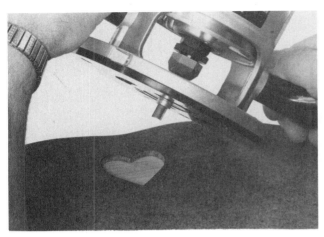

Illus. 13-31. You must carefully tip a fixed-base router, as shown, so that the rotating bit enters the work where desired without accidentally cutting the template.

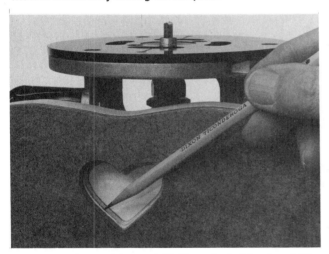

Illus. 13-32. The resulting cut is ³⁄₃₂″ from the inside edge of this template.

Attaching Patterns and Templates to Workpieces

Patterns and templates can be held or secured to the workpiece in a variety of ways. They may be clamped or nailed, as previously illustrated, if nail holes are not objectionable in the finished product. Templates can also be attached with double-faced tape or spot-glued with hot-melt adhesives (Illus. 13-33–13-36). Clamping the template to the work is the easiest method. If the template material can be clamped so the planned cut can still be made without interference, do it (Illus. 13-37 and 13-38). (Refer to Chapter 29 for information about making vac-

uum clamps and templates for router table work that can also be used for many hand-held routing applications.)

The hinged template fixture shown in Illus. 13-36 is another technique that is useful for many jobs. This combination holding device and template is an especially

Illus. 13-33. The single plywood template/pattern shown on the left was used to create all the router-cut designs in the board at the right. The pattern was shifted horizontally and vertically with successive cuts.

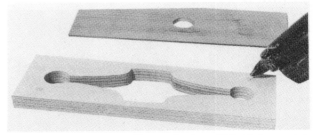

Illus. 13-34. Two spots of hot-melt glue will hold the work to the pattern so that the hanging lamp segments shown in Illus. 13-2 can be pierce-routed. Notice that the work has a pre-bored hole.

Illus. 13-35. A pattern for routing the profiles of wooden switch plates. Short, pointed nails and the downward pressure of the router will hold the work in position during routing.

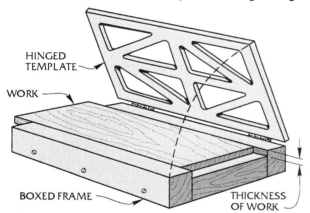

HINGED TEMPLATE

WORK

BOXED FRAME

THICKNESS OF WORK

Illus. 13-36. This hinged-box template can be used to quickly position the work for routing. The work is held secure with pointed nails in the bottom of the jig.

Illus. 13-37. A template designed to cut stock-butt sockets for a gun cabinet. Note that the template's surface is inclined over the work and that a lipped stop locates the template against the edge of the workpiece.

Illus. 13-38. The routed recess cuts have inclined bottoms, to prevent the guns from tipping over.

good system for routing surface designs, for through-routing as in piercing jobs, or for cutting recesses, such as is done to the backs of switch plates or to make shallow boxes and trays.

Inclined or tilted template surfaces make interesting cuts which I seem to find frequent uses for (Illus. 13-37–13-41). Such cuts were used to make the large cutting board illustrated in Illus. 13-39. Inclined templates proved to be the solution to the problem of cutting the very large recess in the cutting board with a portable router, and cutting the bottom surface of the recess flat and on a slant so that meat juices would drain to the back.

A template was obviously needed to cut the large, flat, inclined recess area. Wedges placed under the template automatically made a progressively deeper cut from front to back. The template was either held securely with double-faced tape or temporarily glued down with a few spots of hot-melt glue. The bit used was larger than one used with any commercially available template guide. The router base was also enlarged or extended so that it rode on top of the template throughout the entire recessed-surface-cutting operation.

Shop-Made Template Guides

Shop-made template guides fitted to special shop-made sub-bases for your router will expand your template-routing capabilities. Although not as accurate or as dimensionally uniform as machined-metal commercial template guides, my own innovations have served me well for jobs that are less precise than needed for making dovetails, inlays, setting-in hinges, etc. I use short lengths of plastic pipe and aluminum tubing that I press-

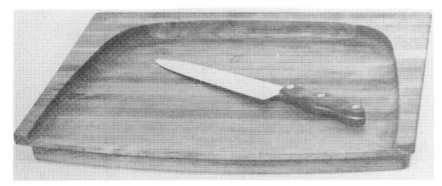

Illus. 13-39. Making this kitchen cutting block with an inclined surface presented some interesting challenges that were essentially solved by utilizing basic template-routing techniques.

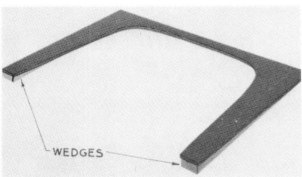

Illus. 13-40. Wedges were placed under the template used to rout the inclined cutting surface of the cutting block shown in Illus. 13-39.

Illus. 13-42. A plastic-pipe template guide glued into a clear-plastic sub-base. A "size compensation" ring has been slipped over the template guide so the cut will be further from the edge of the template.

Illus. 13-41. An extended router base. It has its own integral template guide that is simply a short length of plastic pipe fitted tightly into a bored hole.

fit and/or glue into various router bases (Illus. 13-41 and 13-42). You can also make "size compensators," which are washer-like rings that fit over the barrel of your template guides to change their outside diameters. This allows you to make multiple or parallel irregular cuts using just one stationary template (Illus. 13-42 and 13-43).

Illus. 13-43. A large, washer-like ring that fits over the barrel of a regular template guide permits the routing of two parallel, concentric circle grooves with this one template.

Plunge-Router Boring with Templates

Template-routing is often a neat, quick way to make perfectly vertical holes, in limited sizes, when no drill press is available. The template will locate holes, so no layout is necessary. Illus. 13-44–13-46 show how one template is used to trim a part to its finished size and shape, and then, without being removed, to locate holes that have to be plunge-cut into it. To make the holes, you have to match the outside diameter of a template guide to the locating holes bored into the template. The template guide can be *any size* that has an inside diameter greater than the size of the bit. The barrel of the template guide will thus position the router exactly over where the de-sired hole is to be plunge-bored. *Caution:* Only use router bits, not drills or boring tools, in the router.

Other template-routing or positioning applications will be found in the following chapters. For example, inlaying and setting in hardware are two jobs involving template work. They are discussed in the next chapter.

Illus. 13-44. Trimming and smoothing the outside profile of a workpiece which was first rough-sawn 1/16″ to 1/8″ oversize.

Illus. 13-45 (above left). A template with boring holes of the same outside diameter as the template guide mounted in the plunge router. Illus. 13-46 (above right). Using a plunge router to bore holes located with a template.

14
Inlaying and Setting In Hardware

Inlaying and setting in various kinds of hinges and other hardware require the cutting of shallow recesses or mortises into a surface—a job perfectly suited to the hand-held router. This chapter explores various techniques for inlay work involving irregular or circular designs. Straight-line inlays, mostly used for borders on square or rectangular panels, are generally just parallel grooves which are easily cut equal to the width of the inlay strip and slightly less in depth than its thickness. When used with an edge guide, the router is perfect for strip or line inlay work.

Decorative, curved inlay designs can be cut into solid wood (Illus. 14-1), veneered panels, plastic laminate, hard surface materials, or almost any flat, routable material. Almost all inlay work is best done with a plunge router because the bit enters the work vertically, ensuring a clean, uninterrupted edge.

The easiest and quickest way to produce clean and precise irregularly curved inlay cuts is to guide the router with a carefully made, female-type recessed template that is cut to the required size and design (Illus. 14-2). When used with a special inlay kit accessory for your router (Illus. 14-3), this single template guides the router in making both the recessed female cut into the surface and the male plug or inlay that fits into it. This inlay kit (Illus. 14-3) consists of a template guide that fits to your router base, either a ⅛″ or ¼″ router bit, and a bushing or collar that fits over the business end of the template guide. This bushing has a wall thickness equal to the cutting diameter of the accompanying bit. A ⅛″ bit kit is recommended for inlaying veneer, laminate, and plastic not exceeding ⅛″ in thickness. The ¼″ bit kit is recommended for inlaying material up to approximately ½″ in thickness. One distinct advantage of the smaller-sized kit

Illus. 14-1. Inlays are usually of contrasting colors of wood, to attract the viewer's attention.

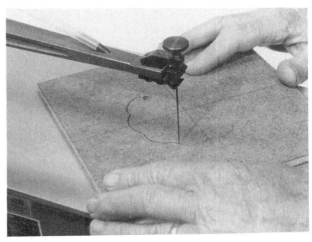

Illus. 14-2. Scroll-sawing a female-type template from ¼″ thick hardboard.

Illus. 14-3. CMT's inlay kit fits Porter-Cable, Black & Decker, and similar router sub-bases. It consists of a solid-brass template guide and bushing and a ⅛″ solid-carbide, downcut spiral bit.

Illus. 14-4. When designing and cutting the template, the smallest-radius curve must not be of a radius that's smaller than the curve of the bushing or collar that fits over the end of the template guide.

is that you can rout more intricate designs. This is because the smallest-radius curve cut into the shape or design of the template must *not* be smaller than the radius of the collar or bushing (Illus. 14-4).

Making an Inlay Template

One-quarter-inch Masonite® or tempered hardboard is a good material for making an inlay template, but high-quality plywood and plastic are better, although more difficult to cut and smooth. The most common style of template is the female, or recessed, template, with which the router and guide set will travel clockwise around the

inside edge to cut the shape of the recess and the inlay. A male, or plug-type, template can also be used for inlay work, but is more difficult to follow with the router and offers a higher chance of error or miscut if not used very carefully. (If a male template is being used, just reverse the use of the collars and the feed direction of the router.) Make sure that the template's curves have a minimum radius of ⁹⁄₃₂″ when using *most* small kits with a ⅛″ diameter bit. The minimum template-curve radius for larger inlay kits may vary.

Plan the size of the inlay design desired and then enlarge it for cutting the template. Cut the template larger all around by a distance that equals the diameter of the bit. For example, if you want to inlay a 4″ diameter circle and you are using a small inlay kit with a ⅛″ bit, the template should be made with a 4¼″ diameter opening. Be sure either to allow some extra material (3–5″) around the perimeter of the template opening to provide support for the router or to add extra stock just for clamping.

Routing Out the Recess

Clamp or secure the template to the workpiece with double-faced tape. Ensure that the template will not slip. Slip the small bushing or collar onto the business end of the template guide. Better-quality collars are pressure-fitted; other collars have to be tightened with a small setscrew. Set the bit's cutting depth (Illus. 14-5) and feed

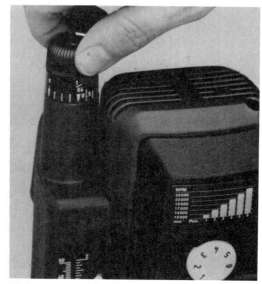

Illus. 14-5. Routers with microfine adjustments are helpful for arriving at precise cutting depths for inlay work as well as when setting in hardware.

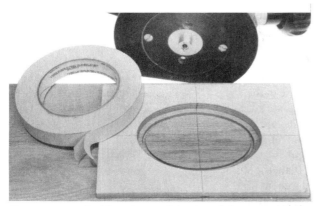

Illus. 14-6. A typical female-type inlay template secured to the work with double-faced tape. Note the cross-reference lines on the template and workpiece used to locate the inlay. Here the initial outline of the inlay's recess has been cut with the collar or bushing on the template guide.

Illus. 14-7. Cleaning up the bottom of the outlined recess with a straight bit.

the router clockwise around the inside of the pattern. The resulting cut will be similar to the one shown in Illus. 14-6. The remaining wood material must be routed away (Illus. 14-7); if it is a laminate material, pry it out.

Making the Inlay

The inlay is made with the same template. Be sure your stock has been prepared to the appropriate thickness. To make the cut, remove the small bushing or sleeve from the template guide. Since you will be making a cut completely through the thickness of the inlay stock, you don't want the inlay piece to shift or kick out when freed. Be sure to stick double-faced tape under the inlay and part of the waste area to hold the work against a flat, scrap supporting backer (Illus. 14-8).

To make this cut, feed the router in a clockwise direction with the router's template guide kept firmly against the inside edges of the template throughout the cut. The resulting inlay piece should fit perfectly into the previously cut recess. Depending upon the size or the nature of the proposed inlay and the kind of materials involved, you may want to cut your male inlay piece(s) first. If you do, the next step would be to outline the recess(es) with the same bit set to the correct depth. Finally, clean out the bottoms with another bit. Cutting the male inlay piece first saves a bit change.

Marquetry Veneer Inlays

Marquetry veneer inlays (Illus. 14-9) are available by mail order in hundreds of dramatic designs, including flower, bird, zodiac, fraternal, and sports designs. These inlays are made of delicately filled small pieces. Most of the

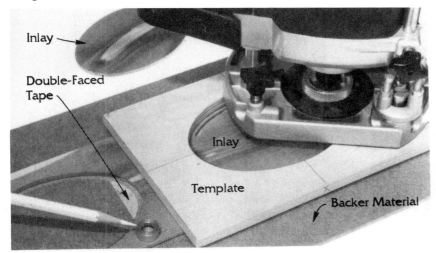

Illus. 14-8. Cutting inlays. Make one continuous cut clockwise, keeping the template guide firmly against the inside edges of the template. Note that the bushing or collar is removed for this cut.

Illus. 14-9. An inlay with a sunburst design adds a professional touch to tabletops, box lids, and similar pieces.

faces of the inlays are taped as a means of holding the parts intact until the inlay is glued in place.

The router can be very helpful in this type of inlay work, which generally involves careful freehand work because making templates is usually impractical. Most decorative veneer marquetry inlays are round-, oval-, square-, or diamond-shaped. The router can't cut square inside corners, but it can be used to assist in the cutting of all of the excess material of the recess with the exception of the inside corners.

The inlay design areas of veneer assemblies often come surrounded with extra backing to protect their edges. This is easier to cut away with a sharp knife than with the router (Illus. 14-10). With the paper face side up, draw a light mark with a pencil to locate the inlay in

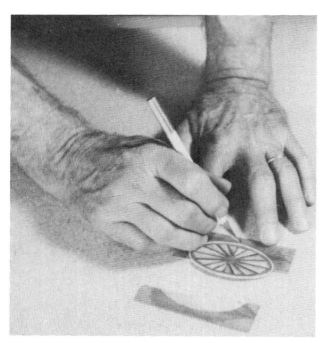

Illus. 14-10. Trimming the waste veneer.

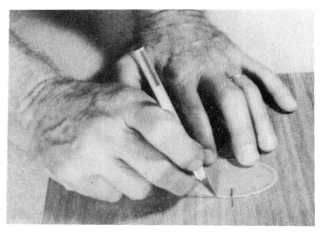

Illus. 14-11. With the inlay's paper side up, make an incision completely around it with a sharp knife. Note the pencil-line reference mark.

relation to the surrounding surface. Next, make an incision completely around the inlay with a sharp knife (Illus. 14-11).

Use an appropriate-size bit. Rout the recess, working from the inside outward almost to the knife-cut incision. Clean up the remaining cut by hand, using the knife to carefully lift out the little waste remaining. Glue the inlay, paper side up, into the recess with yellow glue. Allow the glue to get tacky first (3 to 5 minutes), to reduce squeeze-out. Then press the inlay into the recess, and remove the paper with a moist (not dripping-wet) rag. Do not try to sand the paper off. When the inlaid surface is clean and dry, it is ready for final sanding and finishing.

Setting in Hardware

The router is the perfect tool for cutting the recesses, mortises, and slots required to mount various kinds of hardware. You can use it freehand with purchased or shop-made templates when installing various kinds of hinges, locks, catches, knockdown (KD) fittings, braces, brackets, pulls, etc. Procedures and templates for installing some of this hardware are described below.

Door Butt-Hinge Templates

Door butt-hinge templates (Illus. 14-12) speed and simplify the hanging of doors. You can purchase these templates or make them yourself. Commercial template kits are widely used by full-time, professional builders and contractors. These templates and kits are made in various styles by a number of different manufacturers. Es-

nails hold it securely to the door as it is being routed (Illus. 14-13 and 14-14).

The router is guided by an appropriate-size, base-mounted template guide. The template is not readjusted as it is removed from the door. It is then transferred to the door frame (jamb) to cut the mating recesses, and thus ensure a perfectly accurate alignment and fit every time. The router will cut recesses with rounded corners. These areas will have to be chiselled away by hand if hinges with square corners are being installed.

Those who hang doors occasionally or just want to have templates on hand may find it advantageous to purchase individual 3½" and 4" templates from Sears (Illus. 14-15). When you have just one or a few doors to hang, it's just as fast to trace a line with a pencil around the hinge and rout the recess away freehand, cutting close to the layout line at the proper depth. Then use a sharp chisel to clean up the cut, enlarging the recess precisely to the layout line.

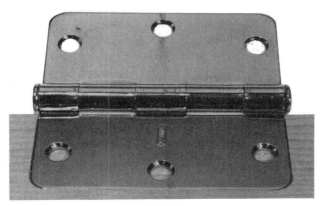

Illus. 14-12. The recess for the leaf of this door butt-hinge was cut entirely with the router guided by a shop-made template.

sentially, they are adjustable templates that permit fast, accurate placement and mortising for installing hinges on doors and door frames (jambs). Once the template has been adjusted for the size and location of hinges, built-in

Illus. 14-13. View showing door-hinge mortising. Note the template guide and straight mortising bit.

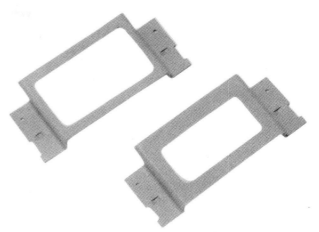

Illus. 14-15. Sears sells individual 3½" and 4" butt-hinge templates. These templates also require a base-mounted template guide (bushing).

However, if you will be mounting many doors, I suggest that you make your own template (Illus. 14-16 and 14-17). Shop-made templates can be designed for any size hinge, and can be cut either with a straight bit and the template guide mounted to the router's sub-base or with a special ball-bearing-guided bit (sometimes called a hinge-mortising bit). This type of bit is also ideal when recess-routing for furniture hardware because the template for it is so easy to make. It's made by cutting the opening to the same size and shape of the hardware being inlaid or recessed. Illus. 14-18 and 14-19 show a shop-

Illus. 14-14. Another view of a door being routed for hinges.

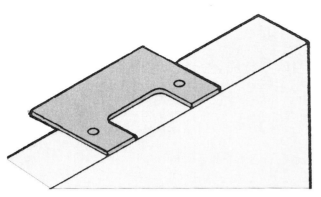

Illus. 14-16. A shop-made hinge-recessing template.

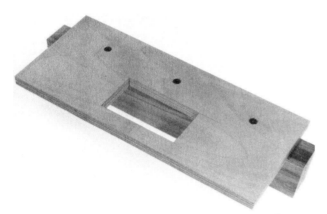

Illus. 14-17. The author's door-hinge template has a clamping block, so nail holes will not show up on the door edge.

Illus. 14-18. Routing a door butt-hinge recess with the template jig clamped to the door. To locate the cuts, the operator will match the centerlines marked on the jig to the corresponding lines drawn on the door and jamb (frame).

made hinge template jig made for routing 3½″ hinge recesses into jambs and doors up to 1¾″ thick.

Mortising for Soss® Hinges

Mortising for Soss hinges is a fairly straightforward process that can be done using one size bit and the edge guide with the router. Both cuts should be made with the router guided from the same door and stile surfaces to ensure that door will be properly flush to the stile.

A plunge router works better, but a fixed-base router will also work. Simply lay out four marks to indicate the start and stop locations for the two-step mortises required (Illus. 14-21).

The manufacturer of Soss Hinges, Universal Industrial Products Co., of Pioneer, Ohio, offers ready-made

Illus. 14-19. Using the same template nailed to the jamb. Reference lines position the template according to the width of the cut required. Design the template with extra stock so that it can be nailed and the door stop will cover the nail holes.

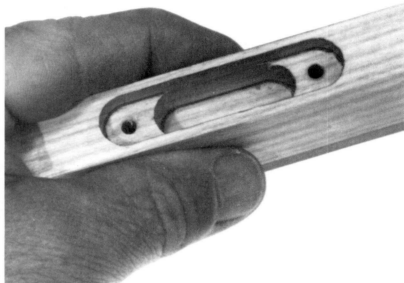

Illus. 14-20 (left). Routing for Soss Invisible Hinges® employs conventional edge-mortising techniques. Illus. 14-21 (above). Identical two-step mortises are cut into the rail (shown) and into the edge of the door. The operator must be more concerned with cutting the shallow mortise depth correctly than the deep one.

templates (Illus. 14-22) for most of its various sizes of hinge. The only adjustment the user must be concerned about is the depth of the shallow mortise. The template is positioned with stops and held to the work (jamb or door) with built-in nails. Two horizontally removable nail-like guide pins locate the length of the deep mortise. When they are removed, the template controls the routing length of the shallow mortise.

Mortising for Knockdown Fittings

Mortising for knockdown fittings is another uncomplicated job for your router. There are many types of knockdown fitting. Illus. 14-23 and 14-24 show a fitting that makes invisible knockdown joints and its application in a right-angle joint.

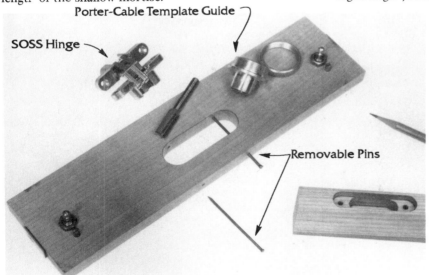

SOSS Hinge

Porter-Cable Template Guide

Removable Pins

Illus. 14-22. A manufactured template designed to locate and control the lengths of the short, deep mortise and the long, shallow mortise for Soss hinge-mortising.

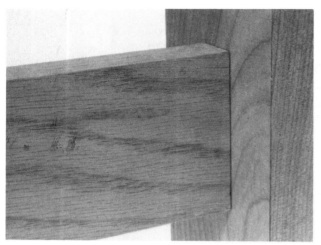 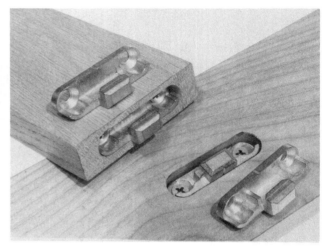

Illus. 14-23. This right-angle joint has an invisible metal connector mortised into each component.

Illus. 14-24. The Knapp Chico connector, available through Häfele America Company, of Archdale, North Carolina, makes invisible knockdown joints for bed frames, railings, table legs, etc. Each mortise is cut the same size and depth because the outside dimensions of each component of the connector are identical.

15
Routing Circles, Arcs, and Ovals

Compass-like jigs can be used to guide the router to produce perfect radiused or oval cuts (Illus. 15-1 and 15-2). Many projects are made that require these types of cut. These include items such as wooden wheels, mirror and picture frames, tabletops, plaques, and home accessories (Illus. 15-3 and 15-4). This chapter presents a few of the many commercially produced jigs and some shop-made ones, to show you some of the many ways this work can be performed.

Circle-Routing Jigs

Commercial Circle-Routing Jigs

Most commercially produced jigs that are not accessories provided by the router manufacturer are designed to be mounted to your router base with screws (Illus. 15-5 and 15-6). The better jigs have adjustable pivot points that are easy to position or to "pinpoint" on the workpiece.

Illus. 15-1. Routing perfect circles with the Dremel Moto-Tool® and a compass-like jig.

Illus. 15-2. Elu's trammel bar being used for circle-cutting.

The desired radius is established by a sliding mechanism and is locked in place with a knob or clamp. The user must drill pivot holes in cheaper or shop-made versions and use a nail for a pivot point.

Some new circle-routing jigs now link the router to the jig with the barrel of a template guide mounted in the router sub-base. With this system, there are no screws holding the router to the jig and the router can be lifted or returned to the jig at will.

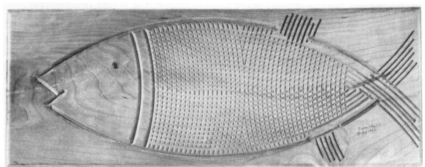

Illus. 15-3. This interesting design, made by Burton Floyd, is based on multiple, true-radiused cuts with V-groove and veining bits worked from four pivot points.

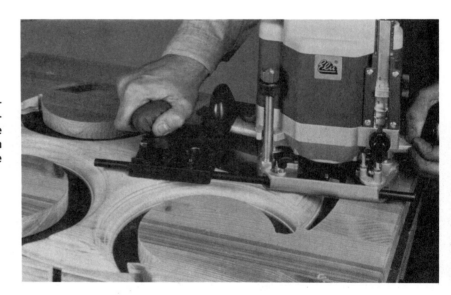

Illus. 15-4. Heavy-duty circle- and arc-routing to produce wood ornamentation. First, the part was cut to its shape with a straight bit. Shown here is an edge-forming operation in which the same center point is being used.

Illus. 15-5. The Lewin router compass jig cuts circles from 8″ to 40″ in diameter, and of even greater diameters with the extender accessory shown. This jig is made of ¼″ clear Lexon® plastic and features a right-angle straight edge that functions much like a router edge-guide accessory.

Illus. 15-6. An Arc® circle jig, made by Porta-Nails, Inc., of Wilmington, North Carolina. It features a sliding-dovetail aluminum construction, an indexing system for repetitive cuts, and a fine adjustment. This jig has a 9″ to 50″ diameter range.

One jig with some interesting innovations is the Circle Mark-N-Cut® jig, developed by Burton Floyd. This jig can be used to draw and rout diameters as small as 1″ up to 44¼″ (Illus. 15-7). Its key component is a flat piece of polycarbonate plastic with a series of ¾″ diameter holes in a patented hole-spacing configuration. The system allows you to set various incremental radii by simply counting the holes. Various kinds of pivoting centers and other inserts provided by the manufacturer or made by the user himself allow the jig to do many craft jobs, including glass-cutting (Illus. 15-8 and 15-9).

Illus. 15-8. The key component of the Circle Mark-N-Cut jig is this piece of plastic with patented hole-spacing configurations.

Illus. 15-9. One of several types of ¾″ diameter flanged bronze bushing can be used to provide the compass action of the jig. One type, shown in the middle of the photograph, has a thin plastic pad that can be secured with tape when no hole in the surface is required. Another type can be secured with a nail. Also shown is a cam center with an offset hole, indicated by the pencil.

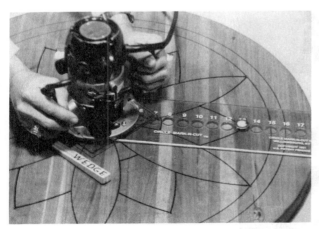

Illus. 15-7. Burton Floyd's Circle Mark-N-Cut® jig in use with a fixed-base router. Note the wedge block; when placed under the jig, it makes it easier to control the entry of the bit into the work surface.

The numbered holes (1–17) are spaced every inch on centers, and the lettered holes are spaced in various fractional increments from the nearest numbered holes. This allows you to use over 100 different radius settings without having to read a ruler. Some intermediate fractional settings must be established when you use the cam center system. In such cases, you have to check your setting with a measurement. With the cam center, any radius is obtainable.

The Mark-N-Cut jig also comes with a square plastic sub-base to attach to your router (Illus. 15-10). This sub-base has a bronze bushing or sleeve to engage any hole selected in the jig. The bit protrudes through the opening in the base, and it should be noted that a ½″ bit is the largest bit that can be used. Porter-Cable or Bosch template guides fit into the guide holes, but you must saw off their lengths in order to use them with this jig.

Illus. 15-10. Bronze bushings serve as the central pivots for both the jig itself and for the square router sub-base which also pivots on top of the jig.

When the manufacturer's bushing or the shortened template guides are used, the router pivots on the jig as the jig itself is rotated to make the cut. This permits the operator to maintain the same hand-holding orientation on the router throughout a full 360-degree feed of the router. The Circle Mark-N-Cut jig is available from Mark-N-Cut Distributors, Inc., Williamsburg, Kentucky.

The Pinske circle compass (Illus. 15-11) is a production jig that features a pivot on a rubber pad with suction grips that clamps itself to the work surface. To activate the rubber pad, simply use the push-button pump on it. The air is pumped out to create a vacuum clamping action. A little lift tab releases the vacuum clamping action. The unit will make circles 13″ to 90″ in diameter. This device is available from The Pinske Edge, Plato, Minnesota.

Illus. 15-11. The Pinske circle compass features a vacuum clamping action.

Shop-Made Circle-Cutting Jigs

Circle-cutting jigs are easy to design and make. They can be very crude (Illus. 15-12) or more refined (Illus. 15-13). They can be made of ¼″ plywood, hardboard, or polycarbonate plastic. Jigs can be designed to be screwed to your router base (Illus. 15-14) or made to link the barrel of a template guide mounted on the router (Illus. 15-15–15-17).

Turn to pages 334 and 335 for plans on making your own circle-cutting jig with a vacuum-clamping central point.

Illus. 15-12. Shown here is one of the most basic circle-cutting jigs: a strip of hardboard attached to the router base with a nail driven through it for the pivot point.

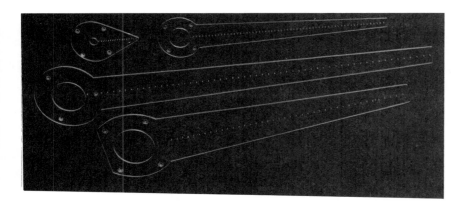

Illus. 15-13. Some circle-cutting jigs made of clear plastic to fit various router bases. The two on the top are designed for trimming routers. The one on the upper left is designed for making small circles such as those used for wooden toy wheels.

Illus. 15-14 (above left). Attaching the circle-cutting jig directly to the router, replacing the sub-base. Illus. 15-15 (above right). This view shows a template guide installed in the base of a plunge router. In use, the barrel or sleeve of the template guide fits into a hole in the jig. The router pivots around on the template guide as the entire assembly is rotated in compass-like fashion.

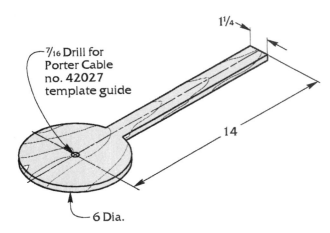

⁷/₁₆ Drill for Porter Cable no. 42027 template guide

1¼

14

6 Dia.

Illus. 15-16. The details for making a circle-cutting jig designed to be used with a template guide mounted in the router base. Note: This type of circle-cutting jig is <u>not</u> intended to be screwed to the base of the router.

Using Circle-Cutting Jigs

Cuts are made either just into the surface or all the way through the stock's thickness. Feed the router clockwise when making a hole and counterclockwise to make circular discs, wheels, etc. When routing a ring (Illus. 15-17), feed the router counterclockwise to cut the inside edge of the hole and clockwise to rout the outside edge. It's also strongly recommended that when you are making through cutouts you secure the workpieces properly on top of some scrap material to prevent cutting into the workbench. Nail or use double-faced tape on the work and the scrap or waste so that no parts shift when the cut frees the piece from the waste.

A Jig That Routs Circles, Ovals, and Arcs

My associate, Carl Roehl, has invented a jig that will soon be marketed that guides the router while it makes perfect circles, arcs, and ovals (Illus. 15-19–15-21). The jig is designed to link the router to the jig with a template guide (Illus. 15-22 and 15-23). This jig may also be used to cut templates for general routing and inlay work. The prototype for this jig can cut circles that range from just 1⅛" in diameter up to 24" in diameter. It can cut ovals that range in size from a minimum of 6" to 10".

Oval-Cutting Jigs

My book *Router Jigs & Techniques* contains plans and instructions for making your own oval-routing jigs. There are two plans, one for using a small trim router (Illus. 15-24) and another for larger routers.

Trend Machinery of Watford, Hertfordshire, England, manufactures a substantial oval-cutting jig designed for Elu routers (Illus. 15-25 and 15-26). It appears that this jig could be modified easily for use with other brands of router. The Trend jig will cut ellipses from around 22" in width by various lengths up to 68" in width by 72" in length, and even larger ellipses, up to 10 feet in the major axis, if the optional trammel bar extenders are used.

Illus. 15-17. Routing rings. Note how the workpiece is nailed on top of a support backer, so when the bit cuts through the thickness of the wood and frees the pieces, they do not shift. The chips in the cut help keep the freed ring from shifting, but using double-faced tape is recommended.

Illus. 15-18. Another quick, easy-to-make circle-cutting jig. Note the dowels set in a wood block with a nail center point. They are 3' or 4' long. This is about as basic a jig as you can make for routing circles to about 8' diameters.

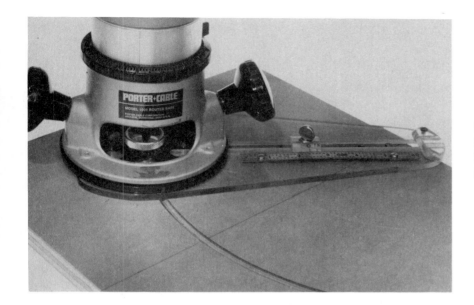

Illus. 15-19. A working prototype of the Roehl circle- and-oval-cutting jig. Here it is cutting an arc. Note the template guide installed in the router's base.

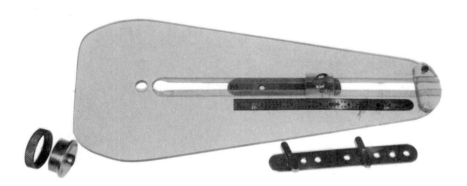

Illus. 15-20. A close look at the essential components of the Roehl circle- and oval-routing jig. The lower right shows two adjustable slides for oval-routing. These are used in ¼" grooves cut at right angles in a base similar to the panel shown in Illus. 15-21.

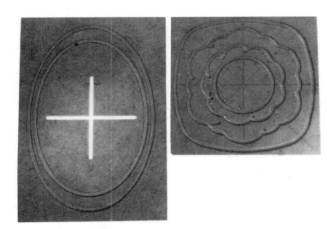

Illus. 15-21. Typical cuts made with the Roehl circle- and oval-routing jig.

Illus. 15-22. A bottom view of the Roehl circle- and oval-routing jig. The barrel of a router-base-mounted template guide fits a hole in the plastic jig. This allows the jig to pivot under the router during the cut.

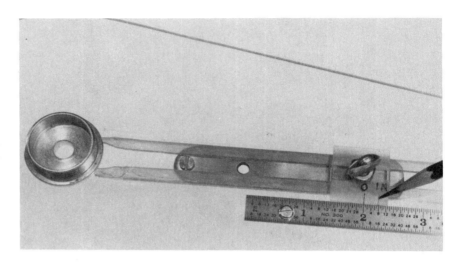

Illus. 15-23. A top view of the Roehl circle- and oval-routing jig showing the template guide in place, the movable slide with pivot point, and the scale and lock plate with two typical measurement marks indicating inside or outside radii when a ¼" diameter bit is used.

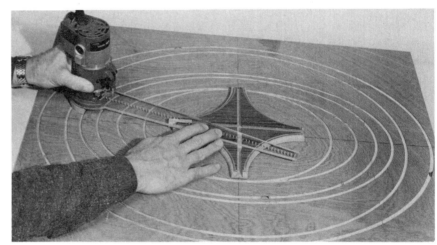

Illus. 15-24. Specific details for making this oval-routing jig and a larger one are given in the book Router Jigs & Techniques.

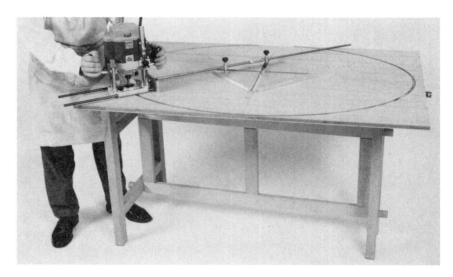

Illus. 15-25. The Trend Ellipse Jig has a ⅝" thick metal-plate base with T-slots for the two movable center sliders.

Illus. 15-26. Using bar extenders to rout a very large oval.

16
Routing Plastic Laminates

Because of their durability, plastic laminates are widely used to cover countertops, edgings and facings of furniture (Illus. 16-1 and 16-2), and cabinets. Laminate materials are very popular and serviceable, and the router craftsman can enjoy considerable financial advantages by working them himself. The process is relatively uncomplicated, and professional results can be obtained immediately with minimal investment in additional tools.

Laminates are also applied to doors, walls, shower stalls, and even windowsills. This material is available in a variety of different colors and in various decorative patterns, such as slate, marble, and wood grains. It can be ordered with a polished high-gloss finish, or in satin, suede, or sculptured designs that have a textured, three-dimensional effect. One of the newest developments by the Formica Corporation is its Colorcore® laminate.

Illus. 16-2. Formica's new Colorcore® eliminates the dark lines typical at joints made with conventional laminates. This corner of a parson's table — a good do-it-yourself router project — exhibits the integral color throughout.

When Colorcore is used in a solid-color pattern, there are none of the dark lines that appear with conventional laminates (Illus. 16-2).

Most general-purpose laminates are a standard ¹⁄₁₆″ thick. V-32 laminate is ¹⁄₃₂″ thick and is used for vertical applications, such as on walls and other surfaces where wear is not as severe. There are other special types of laminate, such as those used with backing or balance sheets and those that are used for cabinet-lining. These laminates are thin, less expensive materials that usually have a plain, nondecorative face.

Core materials (or substrates) to which decorative laminates are applied include particleboard, medium-density fibreboard (MDF), plywood, hardboard, metal, and solid lumber. Solid lumber over 4″ in width is likely to warp. Suitable glues include white and yellow liquids, urea resin, resorcinal, and contact cements. All except

Illus. 16-1. Above: Conventional plastic laminate applied to a panel edge. Below: An edge-to-surface joint. The inexpensive, solid-carbide, ¼″ shank bits shown are excellent for flush- and bevel-trimming. Note the dark seam at the joints.

the contacts require extensive clamping; this makes contact cements the most popular choice for large surfaces.

Basic Process

The basic process involves three easy steps: cutting the plastic laminates, adhering or gluing them, and trimming them. The router is used to cut and trim them (Illus. 16-3 and 16-4). One distinct advantage is that not much is needed in terms of tooling to work with decorative plastic laminates. Small, lightweight routers (Illus. 16-5) are preferred over larger and heavier ones, but any router will do most of the basic cuts. One or two fairly inexpensive

Illus. 16-5. A laminate trimming router is ideal for do-it-yourself plastic laminate fabrication, but any router will do most of the basic cuts.

Illus. 16-3. One good ball-bearing-guided flush-trimming bit is extremely useful for laminate work. If buying the bit just for laminate work, order one with a short cutting-edge length.

bits can do a lot of plastic laminate work. Often, smaller-diameter bits can cut inside corners and other tight areas that cannot be cut with the larger and more expensive bits.

When cutting plastic laminates to cover a particular surface, always leave at least ¼" extra on all overlapping sides and edges. Illus. 16-6 shows a good way to cut laminates to rough sizes with the router. Use a small, piloted carbide bit. Align the line of cut with a straightedge clamped under the laminate, and simply guide the

Illus. 16-4. Any router equipped with a ball-bearing-guided flush-trimming bit enables you to perform this trimming operation—the single most important job in laminate work. Note the minimal exposure of the bit's cutting edge length.

Illus. 16-6. Cutting edge strips of laminate for a clock project or for the vertical edge of a counter or table.

router with the pilot bearing against the straightedge. Plastic laminates are very brittle, so make sure the pieces are well supported during cutting, to prevent them from accidentally cracking or breaking. It's best to cut all the pieces at one time. When making a kitchen counter with laminate on the vertical edge (called a self-edge), it's best to bond that edge first (Illus. 16-7).

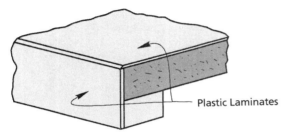

Illus. 16-7. The arrangement of parts for a counter or tabletop with a self-edge. Note that the horizontal laminate overlaps the vertical laminate of the self-edge.

Apply contact cement to both the laminate and the substrate (core) material, and allow both to dry before bonding them. When the two coated surfaces meet or make contact with each other, the bond takes place. Very porous particleboard edges and plywood edges may need a second coat of contact cement. Apply contact cement in accordance with the manufacturer's directions. Apply bonding pressure to the laminates with rollers or by tapping a small block of hardwood with a hammer as you move the block over the laminated surface.

Once the first piece of laminate is applied and pressure-bonded, trimming can begin immediately. Contact cement does not require cure times as do other glues. Trim the self-edge square and flush to the top of the larger surface, using the appropriate bit (Illus. 16-1 and 16-3). Remember, it is essential that solid-carbide or carbide-tipped bits be used (Illus. 16-8). The hardness of the laminate will quickly dull even the best hss bits. Feed directions should be the same as used in conventional edge-routing.

After the edge is trimmed, cut and bond the top piece, overlapping the self-edge in much the same manner as described above. The top can be trimmed with one of the bevel-cutting bits that are pilot- or ball-bearing-guided. These bits put a slight bevel on the edge, reducing the sharpness of the corner. For my own work, I have often used a straight bit (Illus. 16-8), and have then rounded the edge slightly by hand with a fine mill file. Usually, I prefer a very small round-over ball-bearing-guided bit with a 1/16″ radius (Illus. 16-9).

Illus. 16-8. Some straight bits for trimming overhangs. The pilot bit requires a lubricant (such as petroleum jelly) to prevent marring where the pilot bears against the laminate surface. The ball-bearing-guided bit, at right, is a double-cutting production bit.

Illus. 16-9. A small round-over bit with a 1/16″ radius. Other small-radiused bits—called "no-file" laminate trimming bits—are sold by laminate tool suppliers.

If integral-piloted bits are used, they will bear against the laminate during trimming. Use a lubricant such as wax or petroleum jelly to prevent marring. Many professionals prefer this technique because if a ball bearing should freeze, it may seriously score the face laminate. If a ball-bearing-guided trim bit is being used, make sure the bearing is in good shape and that you exert sufficient pressure against it so that it does not spin freely, which could also mar the work.

Specialized Situations

The bare essentials of the plastic-laminate process have just been explored. Following is advice on how to use plastic laminate in more specific situations.

Making Panels with Balanced Construction

One thing to keep in mind when working with plastic laminates is that it is often necessary to make panels with balanced construction to prevent warping. Tabletops, for example, will warp, unless securely fastened all around their edges, like the way a kitchen counter is fastened to the base cabinets. Thus, either apply laminate to both surfaces or use a balance sheet on the underside of the plastic laminate. This material can be purchased from suppliers or directly from cabinet shops (Illus. 16-10 and 16-11).

Illus. 16-10. A balance or backing sheet is essential to keep large, nonfastened panels from warping.

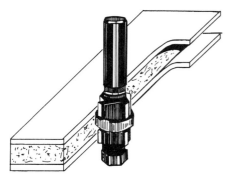

Illus. 16-11. Double-cutting bits are production tools for trimming facing and backing laminates for doors and tables in one pass.

Treating Edges

There are many other ways of treating edges in addition to the self-edge method (Illus. 16-12). Solid wood can be splined or fitted with tongues and grooves flush with the top surface. Then these solid-wood edgings can be edge-formed using any bit-cutting configuration desired. One edge often found on factory-produced laminate products is the plastic (or metal) T moulding (Illus. 16-13). Some mail-order houses now offer T mouldings of various

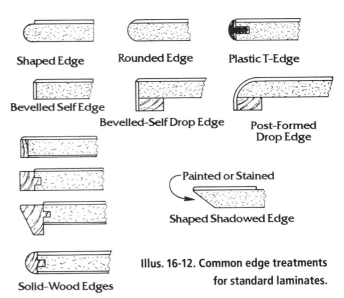

Illus. 16-12. Common edge treatments for standard laminates.

Illus. 16-13. A slot cutter makes the kerf for a T-moulding, to cover a panel edge.

sizes, colors, and wood grains in small quantities for the handyman. Remember, slotting cutters come in a wide range of cutter widths, so it's not much of a problem to set up a T-moulding operation. Be sure to order cutters with carbide-tipped edges that don't dull on particleboard and plywood.

Making an Unusual Raised-Edged Panel

A router, a table saw, and a carbide blade are all that is needed to make the unusual raised-edge panel shown in Illus. 16-14 and 16-15. In this case, the top-surface laminate is applied first in the usual manner. The kerf is sawed with the table saw, and then the wedge is driven in and glued fast with yellow glue. The edges are trimmed flush. Laminate is applied to the vertical edge, and its sharp corners are rounded with a file.

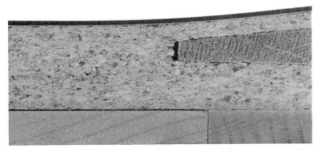

Illus. 16-14. A cutaway view of an actual raised-edge panel.

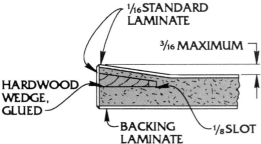

Illus. 16-15. Details for making a raised-edge panel.

Trimming Routers and Accessories

Custom shops, cabinetmakers, and production shops in-volved in the application and trimming of plastic lami-nates use various trimming routers, some with tilting bases and other useful features (Illus. 16-16). Some com-panies provide unusual trimming attachments or special routing accessories and bits (Illus. 16-17) designed for laminate work. Many of these special accessories and jigs are illustrated and discussed in my book *Router Jigs & Techniques*.

Illus. 16-17. Typical small carbide bits used in trim-ming routers. Left: a ⅛″ straight flute. Right: a com-bination flush-trimming and bevel-cutting bit.

Illus. 16-16. Porter-Cable's tilt base for their laminate-trimmer router motor unit.

Illus. 16-18. Trimming an edge overhang flush to the top surface with JCM Industry's special plastic sub-base on a Bosch trimming router.

One noteworthy trimming accessory designed by John Michaels of JCM Industries in North Tonawanda, New York, is made specifically for the Bosch laminate trimmer (Illus. 16-18). It totally eliminates any possibility of a bearing or pilot marking a laminate surface. The jig consists of an adjustable base with a Plexiglas® lip. The plastic edge contacts the guiding surface rather than a pilot or ball bearing (Illus. 16-19).

Seaming Laminates

Sometimes you have to seam or butt-join laminates on large surfaces, such as when mitring kitchen counters or making big L-shaped surfaces. The best way to make and fit such joints is to simultaneously cut both members of the joint with the router guided along a straightedge. The best method is to clamp the two pieces together in the position in which they will be bonded. Make a single,

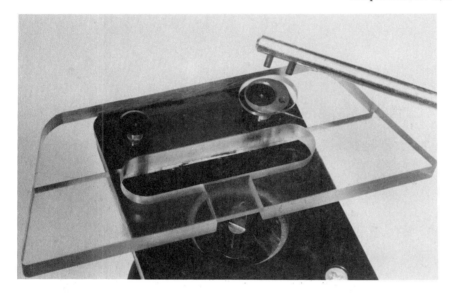

Illus. 16-19. A close-up look at JCM Industry's trimming router plastic base. The adjustable cam and special wrench shown set the edge of the guide relative to the cutting edge of the bit.

Illus. 16-20. The Align-Rite seaming fixture clamps the workpiece and guides a single cut to both edges of the joint in one pass to create a perfect matched butt-joint seam.

uninterrupted cut directly through the joint. You need a fairly good straightedge. Clamp the two pieces onto a flat, scrap panel material. Use a small-diameter solid-carbide bit, set so that it cuts the thickness of the laminate and slightly into the supporting scrap material.

Illus. 16-20 shows a commercially manufactured seaming fixture available from Align-Rite Tool Co., of Tucson, Arizona. This fixture has an aluminum track with quick-action clamps that hold both pieces of the joint securely. Its router guide travels along the tracking fixture and cuts the mating edges in one pass, creating a perfect, matched-cut seam.

17
Routing in the Round

Applying the router to a spinning piece of wood can create round shapes for a variety of decorative or practical uses. Although the router should not be used instead of a turning lathe to regularly create round shapes, this chapter does explore some novel ideas for making light turnings and dowels, and cutting wood threads with a router.

Making Dowels

Almost any router when used with a shop-made or commercial dowel-making jig gives you the ability to create your own dowels (Illus. 17-1 and 17-2). The commercially produced metal jig shown in Illus. 17-3 allows you to make dowels up to 1″ in diameter. The shop-made wooden fixture shown in Illus. 17-4 allows you to make dowels up to 1½″ in diameter. Both jigs involve similar setup and use. The stock is ripped to square cross sec-

tions. A blank is held and rotated with an electric hand drill, which drives and pushes the wood under a core box bit in the router, which is clamped on the jig. When set up

Illus. 17-2. A shop-made dowel-making jig. A square blank is fed under a rotating router bit, with the drive provided by an electric drill.

Illus. 17-3. The Precision Dowel Maker jig is made of welded steel and produces ¼″, ⅜″, ½″, ¾″, and 1″ diameter dowels. It is available directly from S/J Fine Woodworks, 18913 Jackson Drive, Otis Orchards, Washington 99027.

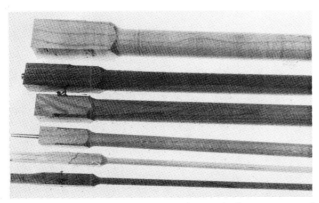

Illus. 17-1. Router-cut dowels can be made from any wood species in all the commonly used sizes.

correctly, both jigs will spin out router-cut dowels with amazing speed and accuracy. The jig shown in Illus. 17-4 is made from three pieces of ¾" northern hard maple that is 5½" wide × 20" long. Baltic birch plywood may be a better choice because it is more dimensionally stable. Two of the pieces are laminated together face to face. This thicker, built-up member becomes the die or exit block. In this piece, holes will be drilled or bored of the same diameter as the prospective dowels.

Illus. 17-5. This close-up shows a square blank at entry and the formed dowel on the exit side. Note the hanger bolt, which is one way of gripping and driving the workpiece in the drill chuck.

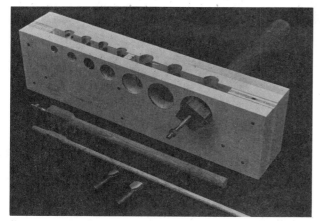

Illus. 17-4. A jig can be made to also produce 1¼" and 1½" diameter dowels.

An entry work-guiding block aids in obtaining a straight starting feed of the square blank into the rotating bit. Bore larger holes into this piece, each at a size that will permit a workpiece with a square cross section to just spin and be fed into this hole. Table 17-1 gives the finished dowel sizes, the large-hole diameters for the entry work-guiding block, and the approximate cross-sectional sizes for ripsawing the square dowel blanks.

The holes in the entry block and the die block must be perfectly aligned to each other on their centers. Illus. 17-5 shows a ¾" plywood spacer inserted between the entry block and the die block. This allows room for chips to fall through during cutting.

As noted previously, Table 17-1 contains approximate sizes for ripping the square blanks. Depending upon hard or soft wood, it is best to saw the squares slightly oversize, so only the very outer corners of the blank will be knocked off as the workpiece spins in the large hole of the entry block (Illus. 17-6). This will take some experimentation to get just right.

One end of the small blank pieces can be sanded or carved round to fit into the chuck of the electric hand drill. Illus. 17-7 shows a ¼" or 5⁄16" hanger bolt screwed into the end of the blank. This bolt is then chucked in a portable electric drill. It's important that the hanger bolt be inserted straight into the blank or the workpiece will wobble, making feeding more difficult. An alternate workpiece-driving system that's perfect when making ¼" and ⅜" diameter dowels is to use the square drives from a standard socket set and/or a socket coupling (Illus. 17-7). This will give you ⅜" and ½" square openings to hold and turn the square dowel blanks. Blanks of larger sizes can

	Finished Dowel Diameter	Large Entry Hole Diameter	Approximate Square Blank Size
	¼" (6 mm)	½" (12 mm)	⅜ × ⅜" (9 × 9 mm)
	⅜" (9 mm)	¾" (18 mm)	½ × ½" (12 × 12 mm)
Table 17-1.	½" (12 mm)	⅞" (21 mm)	⅝ × ⅝" (16 × 16 mm)
	¾" (18 mm)	1¼" (31 mm)	⅞ × ⅞" (21 × 21 mm)
	1" (24 mm)	1⅝" (40 mm)	1⅛ × 1⅛" (29 × 29 mm)
	1¼" (31 mm)	1⅞" (45 mm)	1⅜ × 1⅜" (35 × 35 mm)
	1½" (38 mm)	2⅜" (60 mm)	1⅝ × 1⅝" (40 × 40 mm)

Illus. 17-6. A close-up look at the blank going through the entry side of the Precision Dowel Maker jig. Note that the dowel blank is slightly oversized so that the corners get knocked off.

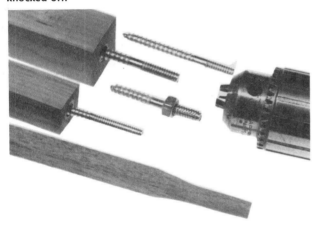

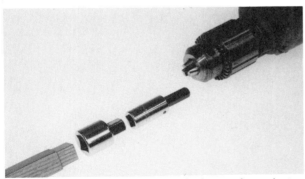

Illus. 17-7. Some drill-chucking methods are shown in top photo. The hanger bolts are used to drive larger-size blanks. Below them is a blank with a sanded round end. Above photo: The square drive from a standard socket set will hold and turn square dowel blanks.

Illus. 17-8. Completing a dowel using the steel Precision Dowel Maker jig.

be machined with short stub tenons ½" square, so they can be driven the same way.

Use a ¾" roundnose bit for cutting dowels up to 1" in diameter, and a 1" roundnose bit to cut the larger-size dowels. Clamp the router on top of the jig and align the vertical axis of the router bit with the front surface of the die block on the entry side.

Next, use the router's own vertical adjustment to set the best depth-of-cut adjustment. This is accomplished through trial and error. The proper bit depth is the most critical adjustment. Look through the entry hole, and adjust the bit so it is tangent to the top of the exit hole. Push a ready-cut dowel through the exit hole, and then bring your bit down so that it just touches it.

When the depth of cut is properly set, the dowel will spin through the die hole. The cut surface of the dowel should come out surprisingly smooth. Its surfaces become slightly burnished from spinning against the inside surfaces of the die hole as the dowel is fed through. If the bit is set too shallow, there will be resistance in the feed and a burned surface will result. If the bit is set too deep, the work will tend to feed too fast and the dowel will have rough spiral grooves on its surface.

The Precision Dowel Maker has an advantage over shop-made jigs in that the wood-to-wood friction that occurs to the shop-made jig will, in time, wear the hole sizes. When the Precision Dowel Maker is used, there obviously isn't wood-to-wood friction.

Router Lathe and Turning Duplicator

This jig (Illus. 17-9 and 17-10) is easy to make and fun to use. It, too, uses an electric drill to spin the wood as it is

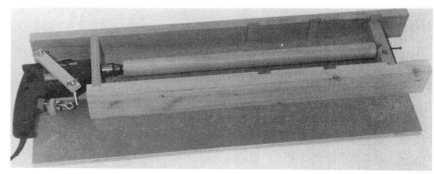

Illus. 17-9. The router lathe and turning duplicator is used to make large dowels, as shown, and to reproduce or copy spindle designs.

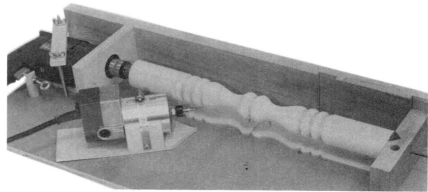

Illus. 17-10. Once the cylinder is made, various contours can be cut with a horizontally mounted router in a sled that follows a thin wood template.

cut with the portable router. This device is intended only for light, short-run jobs, but it could be refined to be more durable. It has a variable-speed control, which is simply an eyebolt turn-screw adjustment against the variable-speed trigger-control switch of the drill (Illus. 17-11).

A square wood blank is chucked to the drill with a ¼″ lag bolt with its head sawn off (Illus. 17-12). Your turning must be of a specific length to fit between the drill chuck and the tail stock. Illus. 17-13, which provides details for making the jig, shows how the tail stock can be moved to any of three dadoes cut into the rails.

The tail stock is crude, but it does work. It's simply a

Illus. 17-11. A close-up of the jig showing its simple speed control. The eyebolt, turned against the variable-speed trigger switch of the drill, controls the rotational speed of the wood under the router.

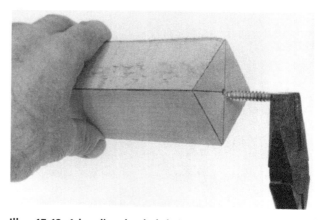

Illus. 17-12. A headless lag bolt being turned into a predrilled pilot hole will be chucked in the drill.

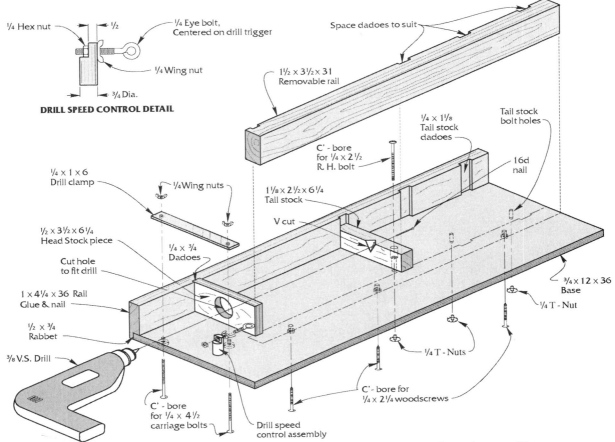

DRILL SPEED CONTROL DETAIL

¼ Hex nut · ½ · ¼ Eye bolt, Centered on drill trigger · ¼ Wing nut · ¾ Dia.

Space dadoes to suit

1½ × 3½ × 31 Removable rail

C' - bore for ¼ × 2½ R. H. bolt

¼ × 1⅛ Tail stock dadoes

Tail stock bolt holes

16d nail

1⅛ × 2½ × 6¼ Tail stock

¼ × 1 × 6 Drill clamp

¼ Wing nuts

V cut

½ × 3½ × 6¼ Head Stock piece

¼ × ¾ Dadoes

Cut hole to fit drill

1 × 4¼ × 36 Rail Glue & nail

½ × ¾ Rabbet

⅜ V.S. Drill

C' - bore for ¼ × 4½ carriage bolts

Drill speed control assembly

¾ × 12 × 36 Base

¼ T - Nut

¼ T - Nuts

C' - bore for ¼ × 2¼ woodscrews

Illus. 17-13. Details for making the router lathe and turning duplicator. Note that only one rail is removable. The tail-stock block can be located wherever matching dadoes are cut in the rails.

block of thick hardwood with a nail pivot. The nail must be placed vertically on a line that runs from the center of the drill chuck and parallel to the surfaces of the rails. If the tail-stock nail is above or below this horizontal line, the turned cylinder becomes a taper. You can turn tapers intentionally by positioning the tail-stock nail at various heights. As shown in Illus. 17-14, I've made a V-cut over the tail-stock center so that I can see the cross marks on the end of the work when I tap in the tail-stock nail.

When rough-turning, begin at a slow speed. Any flat-cutting bit will cut off the corners as the router is fed over the work (Illus. 17-15). Only a slight increase in spindle speed is necessary for further smoothing and contour duplication work.

When duplicating the shapes or contours of objects, you must cut out a thin template of the desired shape with a band saw or scroll saw. The template must be placed

Illus. 17-14. The tail-stock end of the jig. Note the V-cut for sighting the end of the workpiece, and the vertical row of nail holes for taper-turning setups.

Illus. 17-15. Completing a cylinder-turning operation.

against a reference line from the turning that projects vertically down to the surface of the baseboard (Illus. 17-16). You must remove the front-side rail and the turning, in order to tack the template in place for the duplicating work (Illus. 17-17).

You have to use the router horizontally in a shop-made sled (Illus. 17-18) so that the bit's axis is approximately level or slightly below the middle of the turning. Cut the base of the sled incorporating a pointer or template follower sawn to a radius that matches the bit. In Illus. 17-19 the bit and follower have ¼″ diameters, but they can be any size, depending upon the complexity of the template detail and size of the bit selected.

Illus. 17-19 and 17-20 show the turning process involved in duplicating two identical furniture legs from a single spindle.

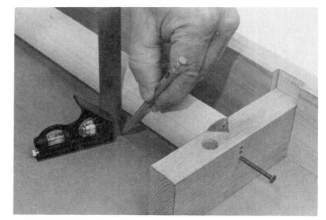

Illus. 17-16. Dropping a line from the turned cylinder to the base of the jig will locate the duplicating template.

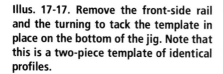

Illus. 17-17. Remove the front-side rail and the turning to tack the template in place on the bottom of the jig. Note that this is a two-piece template of identical profiles.

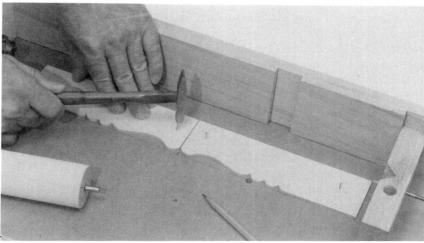

Illus. 17-18. A sled-mounted router with a "pointer" base the same shape as the bit and vertically aligned with it.

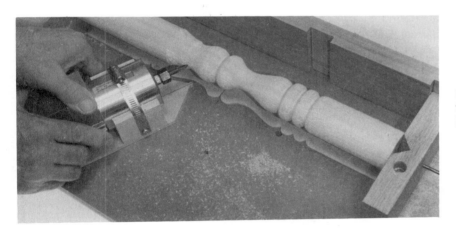

Illus. 17-19. Duplicating the turning profile of the template. Here two legs are being contoured end to end.

Illus. 17-20 (left). When cut apart, the two legs duplicated from a single spindle are identical. Illus. 17-21 (above). A box fixture for cutting flutes in turned cylindrical shapes.

You can incorporate the basic concept involved in using this jig as you customize a jig for your own turning lathe. Fluting and grooving operations can be performed on the router lathe and turning duplicator jig or directly on your own wood lathe.

Lathe-Fluting with the Router

The more complex the shape of the surface to be fluted, the more involved will be the jig or fixture necessary to handle the job.

Fluting lathe-turned work with the router is a fairly common practice among woodworkers. There are two problems to overcome. The first is laying out or indexing the location of the flutes so they end up equally spaced. The second is devising a fixture or support to guide the router to effect the fluting operation.

Some lathes have an indexing lock-pin device on the spindle or spindle pulley, which makes it easy to locate the spacing and hold the work rigid. If your lathe does not have such a device, figure out the flute spacing on a length of flat paper equalling the circumference of the surface to be fluted. Tape the paper pattern to the work, and transfer each flute-cut location with a pencil. You can hold the work rigid and prevent it from turning by clamping the belt to a block of wood held against a pulley.

Illus. 17-21 shows a box fixture set over the lathe that is used to guide the router as it cuts flutes in parallel, straight-turned cylinders. It has two guides to accommodate the diameter of the router base (Illus. 17-22). (A square-base router would be easier to control.) Stops (blocks nailed across from rail to rail) can be used to limit the travel of the router, making all flutes uniform in length.

If the turning is round and also tapered, then the rails supporting the router must be cut to conform to the contour of the desired fluted area. Just about any conceivable fluting job can be handled with a router, if the proper fixtures or router-guided techniques are used.

For more information about machine-like devices that incorporate routers for wood turning, refer to Chapter 33.

Balancing Large Cylinders or Glue-Ups

Use the router with an extended base and a rough-turning frame fixture to help balance large cylinders or glue-ups before mounting them in the lathe. Heavy, unbalanced turnings can be dangerous and may damage your wood lathe if they are not balanced before being turned. Illus. 17-23 shows a technique I use occasionally when turning heavy green chunks of wood. Due to the weight of such turnings, they must be well balanced in advance of actual lathe-turning, to eliminate excessive and often dangerous vibration. The simple fixture used facilitates router removal of protruding knots, unevenness, or other bumps that create out-of-balance turnings. A router with a large bit and extended base is passed over

Illus. 17-22. A close-up view that shows flutes being cut into a turning.

Illus. 17-23. A router with an extended base, as shown at left, is used to true and balance heavy green logs prior to their being mounted in the turning lathe.

the workpiece until it is true. The workpiece is simply hand-rotated on two nail centers at intervals during this routing operation.

Freehand-Carving

With the router motor held stationary, horizontal, and near the edge of a workbench, it is possible to do some wood carving (Illus. 17-24). Suitable metal-cutting burrs, rotary files, or ball mills can be used. *Caution:* Make sure the cutters are designed for high rpm and sized according to the router's horsepower. Do not exceed the speed limits of any tool or carving burr. You may need to use a router speed control to slow the router.

Do not use two-wing or single-flute router bits for wood carving. These could be dangerous, because they may dig into and throw the work. Burrs and files have many cutting edges with small flutes that limit excessively deep cutting, unless forced by the operator. Make sure you know what you're doing and are aware of the limits of your tools and the hazards associated with their use.

Router-Cut Wood Threads

The threading of round dowels can be cut in several ways. Some very complicated jigs with thread box guides can be concocted to cut both internal and external threads. One such user-made device with plans and instructions was published in an article by R. J. Harrigan in the March/April 1983 issue of *Fine Woodworking*. In the same

Illus. 17-24. A horizontally mounted router permits freehand-carving with the appropriate tools. Observe or practise the proper precautions such as using a speed control to reduce speed when necessary.

issue are plans for a router-table wood-thread box by Andrew Henwood. This one cuts external threads. Anyone considering making a wood thread-cutting accessory should refer to these articles.

The Beall Tool Company of Newark, Ohio, sells a router wood threader (Illus. 17-26). It consists of a plastic base fixture that takes interchangeable cast-plastic lead inserts that control the pitch of the router-cut thread. Inserts are available in five sizes of threads in the right-hand configuration and three sizes for left-hand threads. Matching taps are also available, as well as a project book (Illus. 17-27).

Illus. 17-25. A variety of router-cut wood threads made with a Beall jig and the appropriate tooling. See Illus. 17-26.

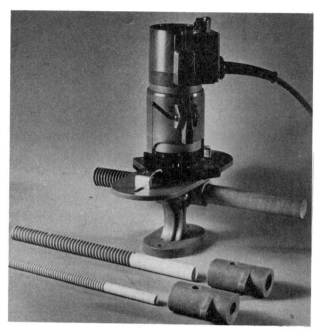

Illus. 17-26. The Beall wood threader cuts several sizes of external threads in both right- and left-hand styles when used with any router and a special three-fluted, 60-degree bit. Note the interchangeable lead inserts shown in the foreground.

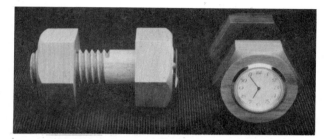

Illus. 17-27. Some clock projects incorporating router-cut wooden threads made with the Beall wood threader.

The Beall wood-threader router attachment will cut external threads in hardwoods and softwoods, as well as some plastics. Any router may be used, even lightweight trimming routers. Small routers actually work best. The Beall attachment is relatively easy to set up. The router is accurately positioned with the aid of an indexing sleeve. The bit cutting depth is set by trial and error, using the router's own depth-adjustment mechanism. Once set, the device is durable enough for continuous production work (Illus. 17-28), with only occasional replacement of the special router bit required. The bit is a hss three-flute, spiral-cut, double-ended veiner bit—and, surprisingly, it is low-priced. A single-end solid-carbide bit is also available, but it's more expensive.

Illus. 17-28. This combination dowel clamp and crank makes it easy to feed the blank dowel. Just turn the dowel into the lead insert, which controls the cutting of the thread pitch.

18
Panel-Routing Aids and Pantographs

In addition to using shop-made templates to rout into flat surfaces, you can also use various panel-routing jigs or accessories and pantographs to copy or reproduce cuts. This chapter looks at a few of these router-guiding devices.

There are several different types of router accessories and machines available that are designed primarily for making decorative cuts in panels (Illus. 18-1). These devices are primarily used by professional cabinetmakers for decorating cabinet ends, drawer fronts, doors, and similar work. The different pieces of equipment illustrated in this section range greatly in their purchase price and work capacity. Most of the devices operate with the workpiece positioned horizontally, but larger panel-routing machines process the panels supported in a vertical position, to save floor space and facilitate the handling of the large panels.

The least expensive units are designed for making decorative grooves in the drawer fronts and doors of kitchen cabinets and similar cabinetry. These employ basic template- and straight-line routing operations. They consist of four adjustable guide rails and inter-

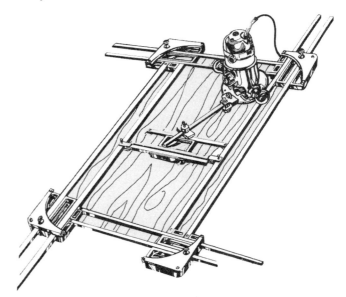

Illus. 18-1. A routed design in a flat panel of Formica's Color-core® surface material.

Illus. 18-2. Porta-Nails, Inc.'s router-panel template, shown set up with an accessory arc attachment.

Illus. 18-3. The Porta-Nails router-panel template is used for door- and drawer-routing.

changeable corner templates of various designs. When these units are used, the router (with a base-mounted template guide) is moved along the rails and corner templates to make the cut.

Porta-Nails Router-Panel Template

The Porta-Nails router-panel template (formerly the Wing template) is shown in Illus. 18-2 and 18-3. It is manufactured of steel tubing and die-cast aluminum components. The unit adjusts to fit any square or rectangular workpiece ranging in size from 3½″ × 3½″ up to 24″ × 36″. Extension bars are available to increase panel-size capacity to 84″ × 84″.

Porta-Nails, Inc., of Wilmington, North Carolina, offers 16 basic border and corner template design sets. Four come with the purchase of the accessory. When the company's various templates are used in combination with each other, a great number of new design configurations become possible. An optional arc design attachment (Illus. 18-4) is also available from this manufacturer.

Illus. 18-4. The arc-design attachment in use.

Bosch Cabinet-Door Template System

The Bosch cabinet-door template system (Illus. 18-5) features a slightly different operating system than the Porta-Nails router panel template. The router base rides directly on the door surface, so no template guide or bushing is required. Bosch sells 12 different corner templates for doors—most are reversible—for routing matching drawer front designs. Also available are arch-topped templates for various-size doors (Illus. 18-6).

Straight-line cutting is accomplished through the carriage-and-rail system. This also permits straight-line grooving (Illus. 18-7), to make vertical or cross-plank designs. The corner templates are made of phenolic plastic. This unit will handle door panels up to 2″ thick, 36″ wide, and 60″ long.

Exac-T-Guide

The Exac-T-Guide (Illus. 18-8) is a straight-line router-controlling device that was actually designed so that the operator could use a circular saw on-site to cut big panels, instead of using a table saw. The unit functions like a huge draftsman's T-square, with a router (or saw) fitted to a shoe that slides along the 52″ extruded aluminum "blade." The "head" moves in another extruded aluminum guide, or U-channel, 52″ or 104″ long, for a lengthwise cutting capacity of 8′ 4″ when used with two U-channels end to end. Its cutting width capacity is anything up to 50″.

In use, the unit is mounted to a flat panel—usually a 4′ × 8′ sheet. This routing accessory can be used to cut dadoes and grooves (Illus. 18-9) into large panels. Adjustable stops control the lengths of cut. The Exac-T-Guide is sold worldwide by Bradpark Industries, Inc., Toronto, Canada.

Illus. 18-5. The Bosch cabinet-door template system. Note that the router rides directly on the surface of the panel.

Illus. 18-6. Some typical door designs made with the Bosch cabinet-door template system.

Illus. 18-7. Straight-line grooving with the Bosch system. Stops on the bar control the location of the grooves.

Illus. 18-8. Making raised-panel door cuts into the surface of medium-density fibreboard (MDF) using the Exac-T-Guide, which is set up with stops to cut in four directions.

Illus. 18-9. The router in a clamped position cuts flutes or parallel grooves as the unit follows in the U-channel guide.

The Trimtramp

The Trimtramp (Illus. 18-10) is another saw-guided device that offers great potential when converted to carry a router. The router can be fed through the clamped workpiece, or the workpiece can be fed against a fence and past a fixed router. The router is mounted to a soleplate that rides in two extruded aluminum guide bars. This assembly travels over a system of fences and adjustable guides that position the work so 90-degree or angular cuts can be made.

The Trimtramp is available in two sizes with 16″ and 20″ maximum stock-width capacities. It comes fully assembled and weighs only 21 pounds. It can be obtained directly from the manufacturer: Trimtramp Ltd., Etobicoke, Ontario, Canada.

Illus. 18-10. The Trimtramp saw guide equipped with a routing kit.

Pantographs

Pantographs are used to make reduced copies of signs, artwork, and thin relief carvings. They do not make same-size copies. Mounting a router to a pantograph, which is nothing more than four rigid arms with movable joints or connections forming a parallelogram, allows you to use the router as a copying tool. This system can be used to trace flat work and even fully three-dimensional sculpted pieces.

Sears Two-Dimensional Pantograph

The Sears two-dimensional pantograph is a very simple router pantograph. As shown in Illus. 18-11, the router is mounted on one of the four connecting arms comprising the basic parallelogram. One corner of the pantograph is the major pivoting point, which is also where the unit is fastened to the workbench. Diagonally opposite this point is the stylus. The stylus is at the end of one extended arm. This extension allows room for seeing and following a paper pattern or stencil with the stylus. Because the stylus is extended, the pantograph does not copy in a size equal to the original. The resulting design cut by the router is smaller than the original pattern. The Sears pantograph reduces to 42, 50, and 58 percent. The

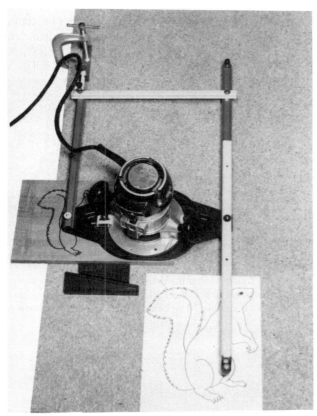

Illus. 18-11. The Sears two-dimensional router pantograph shown with a paper pattern and its reduced copy in wood.

Illus. 18-12. The Sears three-dimensional router pantograph being used in sign work to rout a reduction cut from a larger pattern.

farther away the stylus is located from the arm-pivot joint, the smaller the reproduction will be. To use the pantograph, hold only the stylus arm, not the router. Lifting the stylus arm removes the router bit from the cut. Straight-line cuts are best made by guiding the stylus along a straightedge held securely over the line on the pattern.

The Sears Three-Dimensional Pantograph

The Sears three-dimensional pantograph (Illus. 18-12 and 18-13) can be used for engraving designs of animals, sign letters, and work similarly handled with the pantograph described previously. In addition, this unit will handle three-dimensional carving on a limited basis. It successfully reproduces contoured surfaces of carvings that have a flat back. As with the other pantograph routers, this unit has three settings, which give reduction ratios of 40, 50, and 60 percent.

This unit will fit all routers having bases with diame-ters of up to 6 inches. When used for three-dimensional carving, this machine can only be used to copy patterns up to 1¼″ thick × 13″ wide × 24″ long. Any fairly durable item within these size restrictions can be used for a master pattern.

There are many inexpensive items of plaster, plastic, or cast metal that can be purchased at most hardware and variety outlets, and which make good patterns and nice

Illus. 18-13. Full view of three-dimensional carving with the Sears router pantograph.

products when carved in wood (Illus. 18-13). The process is not fast because small bits, such as a ⅛″ veining bit, must be used to get good, detailed reproduction. The stylus must touch every point on the surface of the pattern.

Hartlauer Pantographs

Walter Hartlauer (85907 Bailey Hill Road, Eugene, Oregon 97405) has been providing the sign industry with large, commercial routing pantographs (Illus. 18-14) for 20 years. Made of aluminum with ball-bearing pivots, two models called the Speed Changer and the Super Changer are available. Neither will make same-size copies. The Speed Changer will only reduce, but the Super Changer will cut reduced and enlarged copies from templates or patterns. Enlarging designs with the Super Changer model is tricky; the manufacturer suggests that it may require the assistance of a helper because the pantograph is so large. The machines can do engraved routing or make cutouts from a pattern. Need-

less to say, there are many pantograph systems for copying and reproduction that can be used for more than sign work.

Illus. 18-14. One of Walter Hartlauer's large router pantographs used for commercial sign work. His machines both enlarge and reduce. Here a number 24 inches high is enlarged to 48 inches.

19
Freehand Routing and Sign Carving

Freehand routing involves any sort of work where the router is guided or controlled exclusively by the eyes and hands of the operator. Skillfully executed freehand router work borders on artistry. The ability to carve sign letters and other exacting cuts in smooth, flowing curves (Illus. 19-1 and 19-2) with quickness and simplicity, like that of an artist making a brush stroke on canvas, does not come overnight. It takes practice to develop the eye-and-hand coordination necessary.

The router artist must also know the working characteristics of the wood materials and the way a particular kind of material will react to the chosen bit and depth of cut. Some woods such as redwood cut like butter with a sharp bit, regardless of grain direction. This is why many professional sign carvers prefer redwood. Douglas fir and some other woods are much more difficult to work freehand because their fibre and grain structure, hardness, or alternating hard and soft growth rings create feeding problems for the beginner. (Chapter 9 examines

feeds and speeds, which are very important factors in freehand-routing success.)

Freehand-routing jobs can be categorized as incised work, raised or relief work, or combination work. Incised work (Illus. 19-1) consists of cutting a line or design into the surface. This class of work can be further categorized as *engraved* or *single stroke*. Engraved designs are recessed areas cut into the surface by making multiple back-and-forth or circular strokes with the router to remove the waste. This kind of work is easiest for the beginner to perfect. Single-stroke work, as the name implies, is cutting irregular or straight lines in just one pass, with the finished cut being uniform in depth and width.

In *raised or relief* work (Illus. 19-2) the resulting design is left untouched and its surrounding background is cut away. This is also a fairly easy process for beginners to perfect. *Combination* work is incised work done within a design that's raised in relief.

There are many situations where using a router free-

Illus. 19-1. Examples of freehand incised routing by Spielman's Cedar Works. Left: Single-stroke line work creates the Viking design. Right: Engraved letters G, B, S, and C are all cut freehand. The straight lines of letters I, B, and H are cut with the aid of a straightedge.

Illus. 19-2. The background of relief work can be routed smooth with a flat-bottom bit, or made textured, as shown above. Here the background was removed with a core-box or round-nose bit, simulating hand-carved texturing.

hand is the only possible practical solution. Being able to freehand-rout fairly well increases the operator's confidence to the point where many templates and patterns can be thrown away. And, for production jobs, it is possible to make templates and patterns freehand. Take the incised profile of the duck shown in Illus. 19-3, for example. Normally, this could be outlined with a template and template guide. If it is necessary to cut a hundred pieces like this, then employ the template-guided tech-

Illus. 19-3. Cutting engraved designs such as this one freehand is good practice for the beginner.

nique. But, suppose it is necessary to make only one such design. It is not profitable, and certainly not as satisfying to go through all of the effort to make and set up the template as it is to make the cut freehand in just a matter of minutes. Projects such as this provide beginners with good practice and experience in handling freehand work. Do not try immediately to carve single-stroke letters for a sign freehand; the results will be disappointing.

When freehand-routing, make sure that you can see the bit and the area around the cut clearly. Remove any obstacles in the way. After all, it is crucial to be able to see the layout line and enough of the line in advance of the cut to anticipate when changes of direction will be coming. Remove the sub-base to open up the viewing area, or use the clear-plastic sub-base described earlier and shown in Illus. 19-4.

Illus. 19-4. Sub-bases for freehand routing. The one at the far left is a factory-made base used for a pattern. The others were made by the author. The best one for general work is the second from the right, which is made of clear plastic. The large one at the far right is used as an extended base to support the router over large cutaway areas.

Next, get into the best possible body position for routing. This is accomplished by clamping the work down onto the workbench well in from the edge. In other words, the work should be secured at a location that requires you to reach for it. This forces you to place your arms on the bench and/or the work (if it's large) (Illus. 19-5). The operator in Illus. 19-6 is in the wrong position. The stool or bench you sit on should be almost level or at a very low angle to the work, and should be positioned so that you bend your neck or back as infrequently as possible.

Practising Freehand Routing

Some practice is essential before undertaking a real job. Prepare workpieces that are relatively free of knots or other defects. Draw some lines on the workpieces in

Illus. 19-5. The position that ensures the best control of the router is with the operator's arms reaching forward and resting on the work surface. A good view of the cutting action is essential, as are safety glasses, a dust mask, and hearing protection.

Illus. 19-6. Poor freehand-routing position for the beginner. The work is too close to the edge of the bench to allow for arm support; this reduces router control.

different directions to the grain. Use a sharp carbide bit no bigger than ⅜″ or ½″ in cutting diameter. An hss bit will dull quickly.

Freehand-routing feeds should be fairly slow for beginners. Set a shallow depth of ⅛″ to ³⁄₁₆″. Practise following along the side of the layout lines as closely as possible. Make a vertical straight cut across the grain. Use a pulling stroke, cutting in a direction from the top edge down towards the lower one (the edge nearest you) (Illus. 19-7). You will notice that the router has a tendency to pull itself to the right. It will be necessary to learn how much to press against the router throughout the cut so that it makes a reasonably straight line. Do the same in routing from right to left with the grain of the wood.

Next, practise angled cuts to the left and to the right obliquely against the grain. Then practise curves. Cut curves to the left and then to the right to get the feel of the router's reaction to grain directions.

All practice cuts begin with a shallow depth and a slow feed rate. There will be some burning of wood when you start practising, but this is to be expected. Increase the speed of feed to the point where you still feel comfortably in control of the router and the burning or smoking ceases. If necessary, set the bit shallower. It is always possible to increase the depth for a second pass. The second, deeper cut will more easily keep the bit following the layout line. Also, in the first practice sessions, attempt to simplify the feed directions.

Make most vertical cuts across the grain with a pulling feed direction, that is, in the direction from the far edge of the board downwards, towards yourself. Make horizontal cuts with the grain, in a right-to-left direction.

With more experience, you will be able to feed the

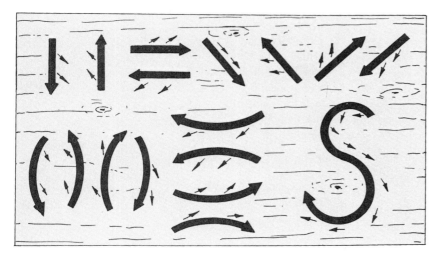

Illus. 19-7. Shown here are strokes you should practise, and the wandering tendencies of the router that will follow when it is fed in various directions in wood with horizontal grain. The large arrows are the intended feed directions. The small arrows indicate the direction the router tends to go if it is not physically restrained.

router in almost any direction and know which way to apply counterpressure so that the router will stay on its intended course regardless of grain direction. Slow the feed when going through or near knots and wild grain. Wide and deep cuts also require slower feeds and more downward pressure applied against the base of the router.

Practising Engraving and Relief Work

Engraving and relief work are done in pretty much the same way. In both types of work, multiple passes are made to widen a routed area to a layout line. Routing the letters for the welcome sign shown in Illus. 19-8 is a good

Illus. 19-8. These engraved letters were routed in multiple passes. The design of the letters made the job easy. The lines of the letters did not have to be perfectly straight, uniform in width, or exactly identical to each other, yet the overall appearance and effect of the sign is good.

Illus. 19-9. Rout the straight portions of letters or other designs using a straightedge guide. The connecting curves are then routed freehand.

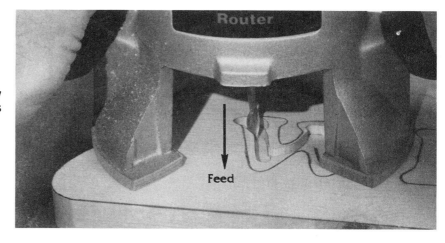

Illus. 19-10. Slowly widen the cut by working on the left side of the previous pass (a counterclockwise feed).

beginner's project. The letter patterns and step-by-step instructions can be found in my book *Router Basics*.

Whenever routing letters or other designs that have any straight-line sections in their profiles, rout these segments first using a straightedge guide, making one pass next to the outline (Illus. 19-9).

To rout out an area such as an engraved letter, begin cutting from the middle of the waste area, working outward to the layout line. Rout away the waste with clockwise circular cuts. This is aggressive cutting, and the router may be difficult to control, but it is all right as long as the routing area is not too small. If the area to be cut away is small and you are nearing the layout line, use a counterclockwise feed (Illus. 19-10 and 19-11).

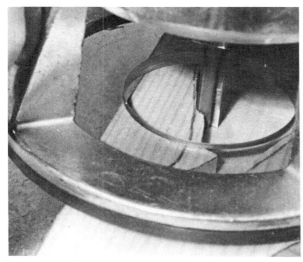

Illus. 19-11. A final trimming cut next to the layout line. Note that this last cut removes a minimum of stock — slightly less than one-half the bit diameter. Also note the feed direction for this finished cut. It is counterclockwise, with the router being pulled towards the operator.

The counterclockwise feed tends to push the router away from the line. A clockwise feed, because of the bit's rotation, tends to grab and pull the router farther horizontally, or beyond the line of cut. Illus. 19-12 and 19-13 show some of the recommended feed directions (which are all counterclockwise) when you are making final passes to the layout lines. Beginners will also find that horizontal cuts are more easily made if they shift the workpiece and reposition it (or themselves) so the cut can be made like a vertical cut with a pull (counterclockwise) stroke (Illus. 19-14).

The completed cutout design should have smooth-flowing curves with the layout lines still remaining. If you slip and go beyond or into the layout line, you can often touch up the cut line. Simply fair out the curve very carefully, using the recommended counterclockwise trimming cut. In most cases, no one will be able to notice the repaired, recut design (Illus. 19-15).

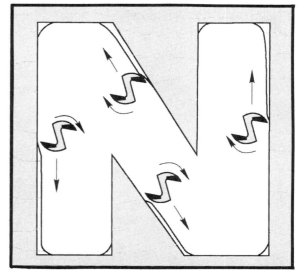

Illus. 19-12. Some of the recommended feed directions for trimming to the line in engraved-letter routing.

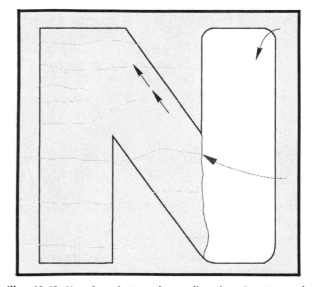

Illus. 19-13. Here is a tip to reduce splintering: Rout towards new wood when possible, not towards or into a previously cut area.

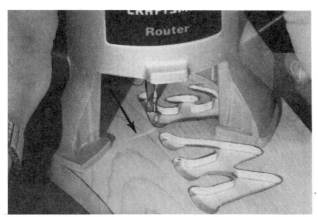

Illus. 19-14. Reclamp the workpiece or position yourself so that you can make horizontal cuts with a pull stroke, such as on the top of the letter E, as shown here.

Illus. 19-15. If you fair out the line with a new trimming cut, as indicated by the pencil, the outline of the letter will appear smooth and flowing again.

After gaining some experience and confidence in handling single-stroke freehand router work, you will be able to simplify or shorten certain procedures. For example, in engraving work you will be able to cut the outlines first, cutting right to the lines in one pass and then hogging out the waste in between. Another approach would be to use a combination of techniques in various situations that allow you to rout deeper and faster.

Illus. 19-16. The freehand-routed design on this picture frame adds an artistic touch.

Practising Single-Stroke Work

Single-stroke work (Illus. 19-16 and 19-17) is best learned by selecting designs on which you do not have to make precise cuts. In other words, if you are routing signs, do not select an alphabet design with perfect straight lines and perfect compass curves. Instead, use one as shown in Illus. 19-17 and 19-18. A technique for laying out sign designs with chalk is shown in Illus. 19-19. Once you have laid out the design in chalk, outline the letter widths in pencil so that you can feed the router with the bit cutting between the lines (Illus. 19-20).

Illus. 19-17. This sign was freehand-routed with the single-stroke technique. A round-nose or core box bit was used.

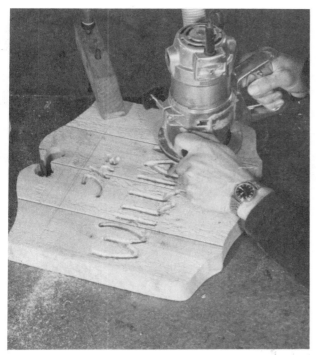

Illus. 19-18. Freehand-routing a name sign using the single-stroke technique.

Illus. 19-19. A chalked layout has been outlined in pencil to a width approximating the bit diameter.

Illus. 19-20. The operator's view of single-stroke freehand routing. Note the use of a roundnose bit for this letter style.

Old English Letters

Old English or modified Old English letters (Illus. 19-21 and 19-22) router-carved into wood are the mark of a master craftsman. A complete alphabet, developed by professional sign carver Tom McIlree, is shown in Illus. 19-22. The first thought of attempting to master this alphabet might cause some uneasiness, but with practice and by breaking letters down to individual strokes, you will discover it is easier than it looks. The best result will be achieved by selecting the easiest wood to rout—unblemished redwood. Choose a few pieces for practice first, since the investment in the wood for the sign itself probably will not be small.

The bit for this work should be an hss or sharp carbide V-bit specially ground to an angle of approximately 60 degrees. A small, lightweight router is the easiest type of

Illus. 19-21. A name engraved in this modified Old English style is dignified and appeals to friends and potential customers.

Illus. 19-22. This complete alphabet routed by Tom McIlree will serve as a guide that you can copy and practise with.

router to use (Illus. 19-23). Most professionals prefer a small-diameter router motor without handles. In fact, one that can be held and controlled comfortably with one hand, such as a trimming router, is even better.

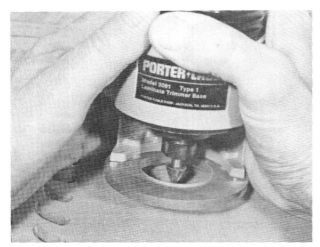

Illus. 19-23. Holding the router ready to make a sweeping serif cut. Note that the router is tipped back, resting on the rear edge of the base. Note, too, that the sub-base is shop-made of clear plastic.

An analysis of the letters shown in Illus. 19-21 and 19-22 reveals that most of the strokes consist of a combination of straight vertical or inclined cuts at various depths. These strokes are ended with sweeping serifs. (A serif is a short line that extends from the end of a letter.) Most serif cuts are made by smoothly lifting the router near the end of the cut so that the lines converge at the

conclusion of the cutting stroke (Illus. 19-23 and 19-24).

Practise making the sweeping, pointed serif cuts first. Straight lines are easy, and Illus. 19-26 shows how to make these cuts.) With the bit set to a suitable depth, lower it into the surface to full depth. Then, with a curved stroke, simultaneously lift the router so the bit cuts more shallowly and move it to the point where it exits from the wood. If necessary, make the first cuts with a layout line marked on the wood, to outline the curved teardrop cut that will be produced. However, with practice, such assistance will not be necessary (Illus. 19-25).

In between making the horizontal and parallel pencil layout guidelines for letter heights, chalk in the letters,

Illus. 19-24. A completed serif practice stroke.

Illus. 19-25. A board of practice strokes made from right to left. After practising these strokes, left-to-right strokes, and sweeping vertical cuts, you are ready to create a sign design.

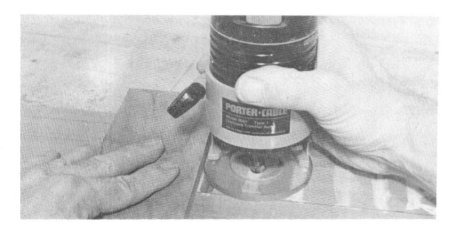

Illus. 19-26. Make all vertical strokes first with the assistance of a small T-square. Try top-to-bottom and then bottom-to-top feed strokes, to determine which is best for you.

Illus. 19-27. The straight strokes are easy. Good layout spacing is important for good-looking results.

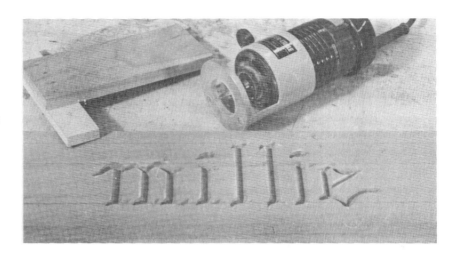

Illus. 19-28. Once all vertical strokes are finished, add the serifs.

copying the shapes of the desired letters from Illus. 19-22 as closely as possible. Use a small shop-made T-square (hand-held, not clamped) and make all vertical cuts with a top-to-bottom feed stroke (Illus. 19-26–19-28). Some sideways pressure will be necessary, because the router will tend to move away from the T-square edge. If this becomes a problem, glue a piece of 60-grit sandpaper under the T-square blade. If it is more comfortable, use a vertical bottom-to-top feed direction for making the vertical letter strokes.

Practice and concentration regarding feed and router resistance in various grain directions are essential to becoming skillful at freehand work. All the effort and practice will be well invested. Very often small jobs—not just sign making, but tasks such as hinge mortise-routing, carving out lock recesses, inlay work, and making cutouts normally made with a jigsaw or scroll saw—can all be handled quickly and accurately, freehand with a router.

My book *Making Wood Signs* covers all aspects of router sign carving and other techniques for making signs. Large commercial signs require heavier, more powerful routers and a variety of special sign-carving bits.

Routing Freehand in a Small Work Area

Illus. 19-29 shows a router connected to a counterweight with a rope and ceiling-mounted pulleys. This is ideal for extensive work in a relatively small routing area where frequent lifting of a heavy router is required. A large hose

Illus. 19-29. This router is suspended over the work from a rope connected to a counterweight by way of ceiling-mounted pulleys.

clamp around the top of the motor unit grips three flat, wire end hangers. The lower tabs of the end hangers are bent over to hook under the hose clamp. It is much less tiring handling the router because of the counterweight. This setup is ideal for continuous-production joint-cutting, edge-forming and freehand work on smaller-sized parts, or intense work in a small working area. The idea can be adapted to incorporate a sliding overhead track to serve larger working areas.

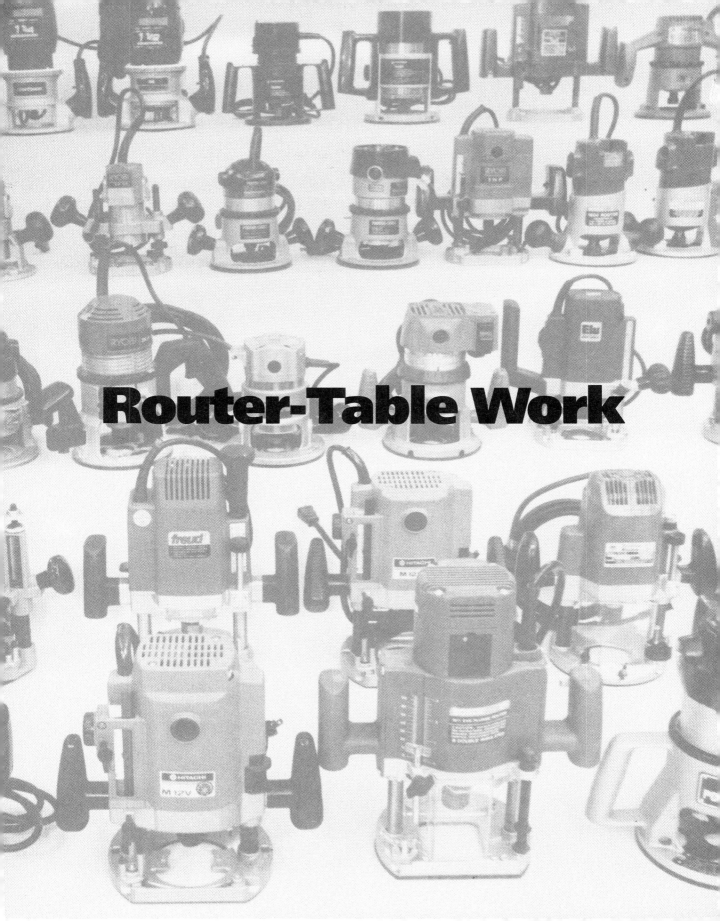

Router-Table Work

20
Commercial Router Tables

A router table is almost a necessity for the serious router craftsman. New routers, bit configurations, and accessories such as precision fences have revolutionized router-table work and make it more fun and more practical than ever before.

A router table is any arrangement where the router (or motor unit alone) is hung upside down under a flat surface with the bit protruding through the surface. In operation, the router is held stationary and the work is passed over or along the rotating bit. Many operations are performed with a fence secured to the table to assist in guiding the work.

With the proper setup, a router table is capable of doing work comparable to that done by a light-industrial spindle shaper. There is an endless list of different operations possible with a router table. Obvious jobs include straight-line dadoing, rabbeting, edge-forming, jointing, smoothing edges, mortising, grooving, slotting, spline cutting, tenoning, and shaping irregular edges. Template reproduction of profiled cutouts can also be done. It is impossible to present every conceivable router-table operation. This subject alone could fill a book. But many of the hand-held router operations discussed in previous chapters can also be accomplished with a router table.

This chapter examines a number of commercially made router tables and offers some selection guidelines. In fact, before you buy a commercial router table, you should review chapters 21–24, which examine shop-made router tables, shop-made accessories, and various routing operations. Custom-made router tables are a better choice than commercial ones, but those too busy to build their own should explore the commercial ones depicted in this chapter. Some of the features and components described in the following pages may provide you with ideas and sources for the fabrication of a table designed specifically for your own needs.

Types of Router Table

There is a very wide range of router tables available that vary in size, weight, price, and quality. Illus. 20-1 shows an all-plastic router table with fence that weighs slightly over one pound. Illus. 20-2 shows a cast-iron router table

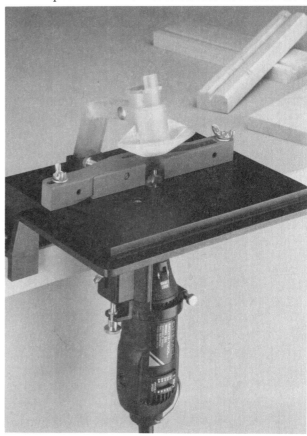

Illus. 20-1. Dremel's small router table has a 6″ × 8″ worktable that clamps to the edge of a table or workbench. This all-plastic accessory is designed for use with Dremel's Moto-Tool.

that weighs about 60 pounds. Many of the lightweight units are designed to sit or stand on top of your workbench. These are fine, except that you will soon find yourself performing router operations that can't be done with these tables. Light-duty router tables are priced so reasonably that anyone can afford one. However, if they don't meet long-term expectations they can be a waste of money.

Be sure to look at the construction features of lightweight router tables closely. Most are made of moulded plastic or stamped sheet metal. Their net weight should give a good clue.

Very light tables may not be durable enough. Lightweight tables are certainly suitable for some individuals' needs. It just depends on the potential use and the overall service expected from them. For anyone planning on only doing light work on small parts, any of the lightweight bench-top models should suffice for some time.

There are also floor-standing router tables available that function like a stationary spindle-shaper machine. However, most are not sufficiently sturdy and need some added reinforcing braces. In order to give you an overview of the array of tables available, I depict and describe in the following sections basic specifications and noteworthy features or limitations of router tables sold by different manufacturers.

Sears Router Tables

Sears router tables vary in style and price. Sears has also recently introduced an "industrial" router table (Illus. 20-3 and 20-4). It is the company's best-quality table, and is essentially a bench-top model. It has a 14″ × 24″ die-cast aluminum table and formed sheet-metal legs,

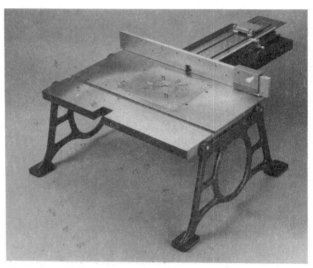

Illus. 20-2. This 65-pound, totally cast-iron bench-top router table by NuCraft Tools, Inc., of Canton, Michigan, has an 18″ × 27″ top on 14″ legs, and many optional accessories. Here it is shown with the Incra jig fence.

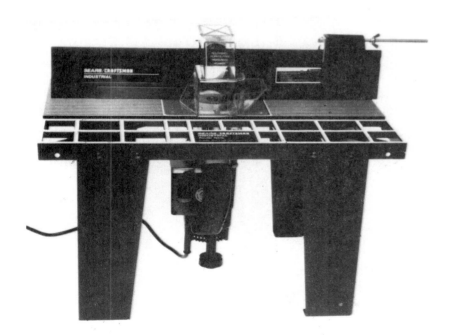

Illus. 20-3. The Sears industrial router table has a 14″ × 24″ surface, weighs 20 pounds, and has optional top extensions and floor legs.

Illus. 20-4. The throat plate for the Sears industrial router table is steel, with levelling adjustments. It will accept most routers.

Illus. 20-5. Jointing an edge on the Sears table. It has an adjustable feature on the out-feed side of the bit. Part of the fence can be set tangent to the cutting circle of a straight bit, so it functions like the out-feed table of a jointer.

but optional to-the-floor legs are available. This table will accept most ½″ collet routers on the market. Its fence has many features, but generally it does not compare in performance to shop-made fences or fences on high-production tables.

The Rebel® Router Table

The Rebel router table (Illus. 20-6) is of solid cast-aluminum construction and measures 17″ high with an 18″ × 24″ table surface. It comes with a ¼″ × 11″ × 11″ insert opening. A split, independently adjustable, 3″ high fence, a mitre gauge, and a guard are standard. The total unit weighs approximately 35 pounds.

Illus. 20-6 (right). The Rebel® router table is of solid cast-aluminum construction and measures 17″ high with an 18″ × 24″ table. (Photo courtesy of Cascade Tools, Inc.)

Porter-Cable Router Table

The Porter-Cable router table (Illus. 20-7) is another bench-top table that measures 10″ high with a 16″ × 18″ surface. It comes with a split, adjustable fence and a built-in switch. This table will accommodate Porter-Cable's larger routers, but the company's catalogue states that "riser blocks must be placed beneath the legs." It has a net weight of 26 pounds.

Illus. 20-8. DeWalt's new router table is a bench-top type similar to the Elu router table. It also comes with an accessory kit consisting of a hold-down guard, clamps, and a template or pattern follower. It is designed only for use with 1¼ hp routers.

Illus. 20-7. Porter-Cable's bench-top router table has a 16″ × 18″ aluminum top and stands 10″ high on formed metal legs.

Elu and DeWalt Router-Table Systems

The Elu and DeWalt router-tables and accessories (Illus. 20-8–20-10) are very similar in appearance, size, and function. The tables are small, approximately 9″ × 10″, with 11″ legs. Sold as a "system," they come with a variety of useful accessories that you don't find with other router tables. Included is a hold-down, a trammel bar for routing discs, and a pattern follower (Illus. 20-9) for copying shapes with curved edges. When the table is clamped vertically to the bench, the router can be used in the horizontal position (Illus. 20-10). In fact, with this arrangement, the horizontally positioned router can be used in a portable mode for edge-forming and other routing jobs, with the attached table and fence providing very accurate guidance. The router table kit weighs 11.7 pounds.

Freud Router Table

The Freud router table (Illus. 20-11 and 20-12) has a hardwood frame and a 21″ × 31″ laminate top. It comes with a drop-in phenolic plate (Illus. 20-13) for mounting

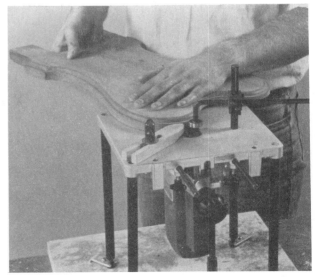

Illus. 20-9. Here a pattern follower is used to duplicate shapes with curved edges on the Elu router-table "system."

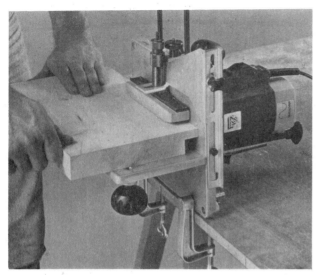

Illus. 20-10. An unusual feature of the DeWalt and Elu tables is that the router can be used horizontally when two clamps provided with the "system" are used to clamp the table vertically to a bench, as shown, so the fence now becomes the horizontal working surface.

Illus. 20-12. It is important that you fasten plywood inside the legs as shown to strengthen and stabilize Freud's leg construction.

Illus. 20-11. Freud's router table is of wood construction, including its fence, mitre-gauge slot, and a spring-loaded hold-down that comes as standard equipment.

Illus. 20-13. The drop-in phenolic plate fits the rabbeted ledge premachined into the top of the Freud table.

your router. A throat insert reduces the bit opening from 3″ in diameter to 1½″. The table stands approximately 3′ high, and is somewhat unstable unless some reinforcement is added. A mitre gauge slot runs along the outside edge, which leaves part of the gauge hanging over the

Illus. 20-14. Freud's unusual mitre-gauge slot runs near the outside edge.

edge (Illus. 20-14). The fence is made of wood. It is flat, nonadjustable, and not very substantial. The table weighs 45 pounds.

Hartville Tool & Supply Router Table

The Hartville Tool & Supply router table (Illus. 20-15–20-17) has a 1″ medium-density-fibreboard top that's laminated on both sides. The mounting plate (Illus.

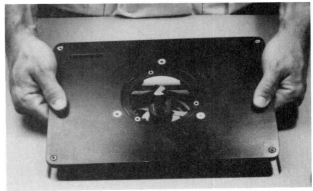

Illus. 20-16. Hartville Tool & Supply's removable router-mounting plate has levelling screws in each corner.

Illus. 20-15. Hartville Tool & Supply's router table has a 24″ × 32″ top, a 7¾″ × 11″ router-plate insert, and stands 31″ high on a mortised-and-tenoned hardwood stand.

20-16) has levelling screws and a replaceable blank insert that can be custom-bored or -sized for any desired bit opening. The stand is mortised-and-tenoned hardwood. Extruded aluminum is used for the fence, which has a vacuum outlet (Illus. 20-17). It weighs approximately 57 pounds. Other accessories designed for the table are also available from Hartville Tool & Supply, Hartville, Ohio.

NuCraft Tools Industrial Router Table

NuCraft Tools, located in Canton, Michigan, produces several industrial router tables. Model 2005, shown in Illus. 20-18, is 24″ × 32″ with a 1³⁄₁₆″ thick laminate top that stands 34″ high on rigid steel legs with a shelf assembly. Note that its router-mounting plate is not centered in the top, but rather it is offset. This feature provides a large in-feed or out-feed area, depending

Illus. 20-17. A back view of the 34″ long extruded aluminum fence that is standard on the Hartville Tool & Supply's router table.

Illus. 20-18. The NuCraft Tools model 2005 industrial router table features metal legs and an uncentered router mounting-plate insert.

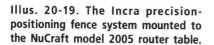

Illus. 20-19. The Incra precision-positioning fence system mounted to the NuCraft model 2005 router table.

upon which side of the table you are working on. It also accommodates precision positioning fences such as the Incra (Illus. 20-19) and JoinTECH fences. Also note that this table has no fence or mitre-gauge slot; this type of table is actually preferred by some craftsmen.

NuCraft makes a model 2000 router table which is similar to the model 2005 table, but has a centered mounting-plate insert opening. Illus. 20-20 shows Nu-Craft's heavy-steel-plate router-mounting plate insert, which is interchangeable with all NuCraft router tables. The company also offers a model 3000 router table, which is similar, but larger than the company's other tables. It has a 32″ × 48″ top. Prices and weights vary from model to model. NuCraft Tools also makes a full line of heavy-duty fully cast-iron router tables and shaper style fences.

Woodhaven Router Table Tops

Woodhaven router table tops (Illus. 20-21 and 20-22) are available in a variety of sizes from 19″ × 24″ to 32″ × 48″, and with various centered or offset mounting-plate configurations. All the router table tops are made of laminated 1″ medium-density-fibreboard

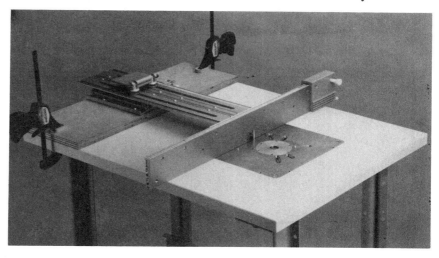

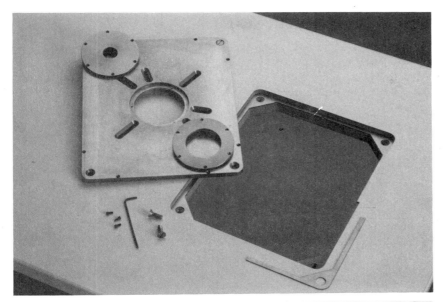

Illus. 20-20. A close-up look at Nu-Craft's router-mounting plate insert. It's made of ⅜" steel plate, is 9" wide × 11" long, and has levelling and permanent mounting screws. It features replaceable shaper-style bit clearance rings that also have levelling screws.

Illus. 20-21. The Woodhaven router table top and drop-in router base. Note that this wall-mounted installation is supported on folding metal brackets to conserve shop (garage) space.

Illus. 20-22. The Woodhaven router table top, folded down and out of the way.

with T-edge mouldings. Clear acrylic plastic that is ⅜″ thick or phenolic router-mounting inserts of various sizes are available, as are several leg- or top-supporting systems, including various hardware for doing-it-yourself. Fences (Illus. 20-23 and 20-24) and other individual router-table accessories are found in various mail-order catalogues or may be ordered directly from Woodhaven of Davenport, Iowa.

RBI Router Table

The RBI router table (Illus. 20-25) is a very basic design. It has a medium-density-fibreboard laminated top that is 18″ × 30″, metal legs, and a one-piece formed sheet-steel fence. The drop-in router-mounting plate is made of ¼″ Plexiglas®. It has a net weight of approximately 48 pounds, and it is manufactured in Harrisonville, Missouri, by RB Industries.

Shopsmith Table-Routing System

Shopsmith sells a variety of different table-routing systems, including a couple that incorporate conventional floor-standing router tables (Illus. 20-26). One unit, called a freestanding table, measures 18″ × 30″ and has a drop-in router-mounting plate.

Illus. 20-27 shows a routing setup on the Shopsmith Mark V 500 multiple-function machine. Shopsmith also manufactures an individual routing-machine station "system" in which you can use the router in its under-the-table setup or you can mount it to an overarm extending from the vertical column of the table (Illus. 20-28 and 20-29). The table is 18″ × 30″ and comes with a variety of accessories, including a dust pickup, a fence, and a featherboard. Shopsmith products are manufactured in Dayton, Ohio. (Turn to Chapter 30 for more information about Shopsmith's overarm features).

Illus. 20-23. The rear side of a Woodhaven router-table fence system is built around a 3″ × 3″ aluminum angle.

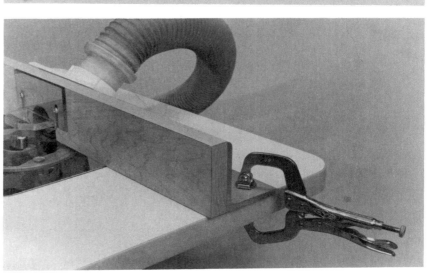

Illus. 20-24. Vice Grip® clamps with swivel pads provide quick adjustments.

Illus. 20-25. The RBI router table features a one-piece sheet-steel fence and plastic drop-in router-mounting plate.

Illus. 20-26. The Shopsmith freestanding router table comes with a variety of accessories, including a dust hookup, a featherboard, and an adjustable fence.

Illus. 20-27. Table-routing (panel-raising) on the Shopsmith Mark V System, which is equipped with a speed-increaser accessory. Note the shaper-style fence, and the guard and dust pickup.

Illus. 20-28. The Shopsmith combination router in two different positions: under the table and mounted to a drop-in base, as shown here, or in an arm bracket extending from the back column for overarm pin-routing applications.

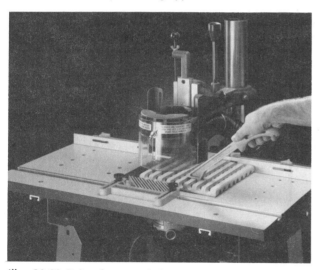

Illus. 20-29. Using the router in its under-the-table mode. Note the overarm guard, dust-extraction system, and the featherboard in the mitre-gauge table slot.

Woodworkers Supply Drill-Press Pin-Router Table

The Woodworkers Supply drill-press pin-router table (Illus. 20-30–20-34) has an innovative design that utilizes the precise features of a floor-model drill press. Made to be utilized with most ½″ routers, this accessory has an 18″ × 24″ table with a fairly small (8¼″ × 8½″) router-mounting plate that may make it difficult to drop in some routers. However, this table has pin-routing capabilities, which many router tables do not.

The router is clamped with a table-adjustment provision that aligns the router's collet perfectly to the drill-press chuck (Illus. 20-33). This arrangement makes it easy to do various router duplicating and pattern-copying work (Illus. 20-34). For one-to-one copies, the guide pin and cutting diameter of the bit need to be equal. Contact Woodworkers Supply Inc., Albuquerque, New Mexico, for more details.

Illus. 20-30. This drill-press pin-router table by Woodworkers Supply, Inc., features an 18″ × 24″ working table. Using the chuck-and-quill assembly of the drill press, the unit performs many duplicating and template-routing operations.

Illus. 20-31. This cast-aluminum frame supports the table and clamps to the column.

Illus. 20-32. The drill-press routing table equipped with an adjustable split fence and a dust-extraction system.

Illus. 20-33. A table-adjustment feature shifts the router so its collet is perfectly aligned with the drill-press chuck. Check this adjustment by attaching a piece of drill rod or dowel to the chuck and lowering it to the collet of the router.

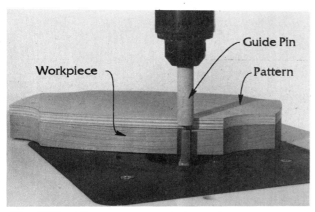

Illus. 20-34. Pattern-routing with the guide pin aligned with the collet and bit.

Rousseau Router Table

The Rousseau router table (Illus. 20-35) is a portable table intended for on-site use by finish carpenters and cabinetmakers. It is a typical medium-density-fibreboard and laminate table, with folding square, tubular legs. It is 20″ × 24″. The basic unit can be fitted with connecting roller supports on the in-feed and out-feed sides.

The Rousseau fence and router-mounting plate are well designed and have some noteworthy features. The mounting-plate table insert measures 9″ × 11¾″ and has snap-in reducer rings (Illus. 20-36). The fence is cast aluminum, and can be adjusted easily (Illus. 20-37). The Rousseau Company is located in Clarkston, Washington.

Illus. 20-35. The Rousseau folding, portable router table with an out-feed, connecting roller support.

Illus. 20-36. The Rousseau router-mounting plate can also be fitted with a Porter-Cable template guide.

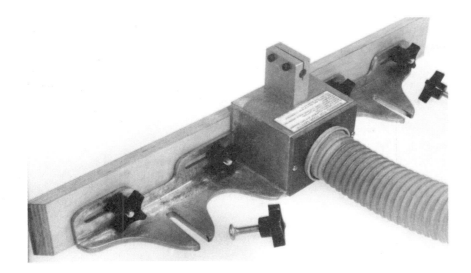

Illus. 20-37. The Rousseau cast-aluminum fence. Various hold-downs, guards, and feed rollers can be attached to the top of the dust-collecting cavity. See Illus. 20-35.

Porta-Nails Router System

The Porta-Nails Inc. "Three-In-One" Router System (Illus. 20-38 and 20-39) has a cast-aluminum top with a standard $\frac{3}{8}'' \times \frac{3}{4}''$ mitre-gauge slot. A provision also allows you to mount the router horizontally with a lead-screw vertical-adjustment mechanism. The overall design configuration permits conventional table-routing work. Also, pin- and pattern-controlled routing is possi-

Illus. 20-39. The third routing mode of the Porta-Nails "Three-In-One" Router System is pin-routing with this optional arm attachment.

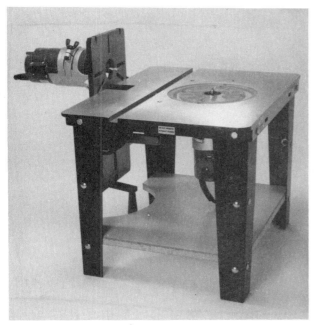

Illus. 20-38. The Porta-Nails router system shown in a standard routing or horizontal routing mode.

ble with the overarm attachment that holds assorted-size pins, centered and aligned, to the router bit. The arm and horizontal router-mounting mechanism is lead-screw-driven and raised or lowered $\frac{1}{16}''$ with one full turn. The manufacturer, Porta-Nails, Inc., of Wilmington, North Carolina, states that this equipment can be used with large, 3 hp routers under the table, routers with up to $1\frac{1}{2}$ hp horizontally. The table stands on just

18″ legs and weighs, without any router(s), only 24 pounds. Various accessories and options such as an air-cylinder pin control, a vacuum chuck, speed controls, and a fence with dust-collection features are also available.

NuCraft Tools Cast-Iron Router Table

The NuCraft Tools cast-iron router table (Illus. 20-40 and 20-41) consists of a 65-pound, 18″ × 27″ top. It sits on two cast-iron, 14″ high legs, and to date it is the heaviest tabletop on the market. This table utilizes the same heavy-duty ⅜″ × 9″ × 11″ steel router-mounting plate (Illus. 20-20) that is used with NuCraft Tools' other router tables.

The table (Illus. 20-40) has two clamping "wings" for mounting shop-made fences or other fixtures. NuCraft Tools of Canton, Michigan, has a variety of attachments already developed, such as straight-line and shaper-style adjustable fences (Illus. 20-41) and mounting extensions for Incra and JoinTECH fences. Soon to be added are horizontal- and pin-routing attachments.

The sizes of tabletops and extensions available from the manufacturer allow their use as table-saw extensions (Illus. 20-41). The threaded cast bosses on the underside accommodate ¾″ pipe legs for added support. The Nu-Craft router tables and accessories are all heavy duty and very well made—especially when compared to the typi-

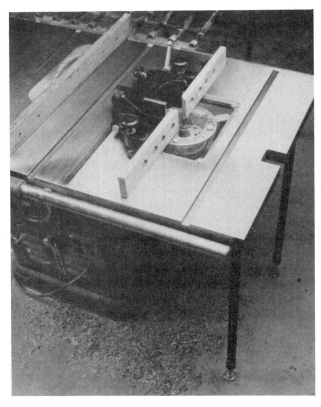

Illus. 20-41. NuCraft Tools' cast-iron router tabletop used as a table-saw extension. Also shown here is the company's shaper-style cast-iron fence.

cal plastic and sheet-metal components of products that currently comprise the market.

Table-Saw-Extension Router Tables

Mounting routers in saw-table extensions offers the advantage of using the saw fence control-and-clamping mechanism and the saw table as an increased work surface. In addition to the cast-iron extension table made by NuCraft Tools (Illus. 20-41), several companies make 27″ standard-leaf extensions of other materials. Bosch makes one of sheet steel.

Mule Cabinet Machine, Inc., Aurora, Ontario, Canada, makes a 15½″ × 27″ laminate-covered medium-density-fibreboard extension table that comes with two formed-steel support brackets (Illus. 20-42). The Mule routing table leaf is available in the United States from CMT Tools, Tampa, Florida.

The Excalibur Machine and Tool Co., Scarborough, Ontario, Canada, produces another table-saw-extension router table (Illus. 20-43 and Illus. 20-44).

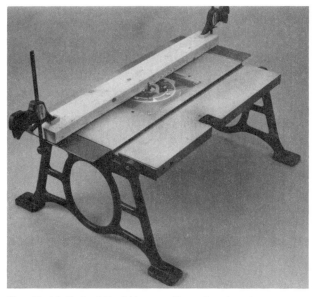

Illus. 20-40. NuCraft Tools' heavy, all cast-iron bench-top router table with a simple board fence that demonstrates the two clamping "wings" (ledges).

Illus. 20-42. The Mule router table extension leaf. It features a router mounting plate insert with three reducing rings providing three different openings of up to 3¾" diameter, the biggest for large raised-panel bits. The router plate insert is 7⅞" × 11⅞".

Illus. 20-43. Excalibur's router-table routing kit and fence brackets. Note the adjustable split in-feed out-feed fence that works off the Excalibur T-slot table-saw fence. Also shown in the setup are the Shophelper® anti-kickback stock-feeding and hold-down wheels.

Illus. 20-44. Excalibur's router-mounting plate is a hinged steel plate that is 12" × 18". It will accept most round-based routers.

21
Shop-Made Router Tables

In recent years, major woodworking magazines and mail-order catalogues have offered "revolutionary" router table plans, each touted as the best. The "best" router table is the one that fully satisfies all your woodworking expectations and needs. Everyone's needs are different. You may need a router table for a wide range of routing operations, one designed for use in limited space, or one just to accommodate a particular accessory such as the precision positioning devices described in the subsequent chapters.

This chapter examines different types of router tables and examines the components of router tables. Use the ideas presented here and in Chapters 20 and 24 to design a router table best suited for your specific needs.

Basic Router Table

A very basic table will serve the needs of many. The one-leg inexpensive basic router table shown in Illus. 21-1 and 21-4–21-6 is intended particularly for new router owners, for those who do not have much workshop space, and for those who want a portable table or a supplement to their primary router table. It consists of two pieces of material 18″ × 18″. The top piece should be ¼″ thick hardboard, hardwood plywood, or plastic. The support-

Illus. 21-1. This one-leg router table clamped to your bench is about as basic as you can get.

Illus. 21-2. The author's Ultimate Router Table designed for a broad range of routing jobs. Plans and construction information can be found in Router Jigs & Techniques. Note the side-mounted switch and speed control.

Illus. 21-3. This horizontal joint-making attachment is just one of many shop-made accessories for the author's Ultimate Router Table.

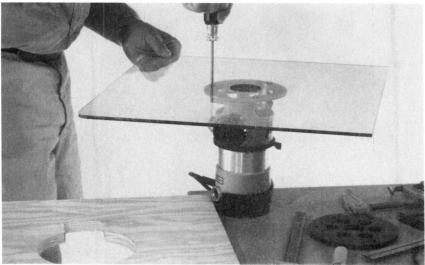

Illus. 21-4. Attach a ¼″ thick panel (18″ × 18″ in this example), with a pre-cut bit hole, directly to the router. This panel will replace the sub-base.

Illus. 21-5. The thick support panel with cut-out router opening rests on the 2 × 4 leg (with dowels) and the edge of the workbench.

Illus. 21-6. Clamps keep the panels from shifting and hold the table to the work-bench. Note that the board (a straight 2 × 4 piece) makes a good fence.

ing piece should be strong and flat and about ¾" thick. The router table is not designed for heavy-duty routers.

Note that the router opening hole of the router table is cut or sized closely to the router, for maximum support. Two clamps and the fence clamps keep the table rigid. The leg (2 × 4) is cut equal to the bench-top height. Two short dowels set into the end match two holes drilled in the thick support piece.

Illus. 21-7 gives the basic design for a light-duty, bench-top router table. Use a ½" thick plywood top for small routers. Use ¾" thick material and modify the plan size for larger routers.

Router Table Components
Tabletops

Most serious table builders make router tabletops from flat sheet material such as particleboard, medium-density fibreboard, or hardwood plywood. Many like the idea of applying plastic laminate to the top, but if you do, also apply it to the bottom surface to balance the panel. This will greatly minimize, but may not prevent, warpage.

How far from the floor should a router tabletop be? One expert says your wrist should be level with the tabletop when you use it. Another expert says your elbow should be level with the tabletop. My suggestion is some-where in between—32" to 35".

The top of a router table used for heavy-duty routers should be attached to sturdy legs or to a custom-designed

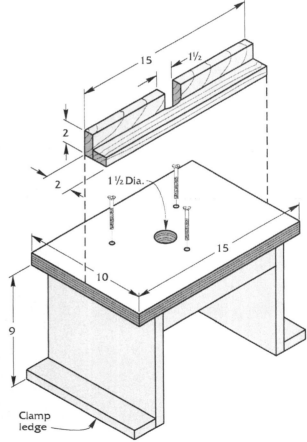

Illus. 21-7. A simple bench-top router table. Modify the sizes to suit your needs.

supporting cabinet with doors, drawers, and shelves for storage. Do not enclose the router without providing a vent or air supply, to prevent the router from overheating.

Mounting Plates

The "drop-in" router plate (Illus. 21-8) is one of the most popular router-table-mounting systems, The thin plate allows you to utilize the full depth-of-cut provision of your router, which is otherwise lessened with a thick top.

The router is simply attached to a rectangular base of a thin, stiff material. It is inverted and held in the table by gravity. The mounting plate is supported on a rabbeted lip around a sawn or router-cut opening in the table.

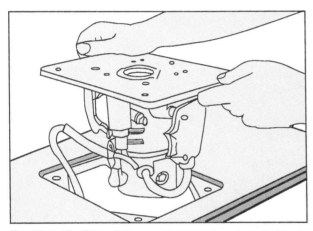

Illus. 21-8. The "drop-in" type of router plate is probably the most popular. The router attached to a flat plate hangs on a rabbeted ledge cut into the tabletop.

Make your mounting plate as small and narrow as possible, even though it may be more difficult to drop into the table. A narrow plate will be stiffer. Place the screw-mounting holes as close as possible to the outside edges of the mounting plate. This will minimize sagging from the weight of the router or the downwards pressure that

is forced against it when you are making heavy cuts such as raised panel work. You may also want to bolt the plate down in each corner for a more permanent installation. I actually prefer bolting it down at each corner, to ensure that there is no movement sideways or up and down. Also, sawdust can work itself between the lip and plate if the plate is just held there by gravity.

Authorities differ as to what is the "best" router-plate material (Illus. 21-9). Each material has advantages and disadvantages. I have found that polycarbonate plastic is easy to machine and is tough. Other individuals recommend acrylic and phenolic plastics, aluminum, and steel, all for various reasons.

Illus. 21-9. Router mounting-plate materials. At left is phenolic plastic. At right is polycarbonate clear-plastic.

High-density moulded-foam inserts are also available. Check all such inserts for flatness (Illus. 21-10). Simply mark, drill, and countersink screw holes for mounting (Illus. 21-11 and 21-12). Orientate the router prior to drilling the mounting holes, to ensure that when it is mounted in the table its switch, depth-of-cut adjust-

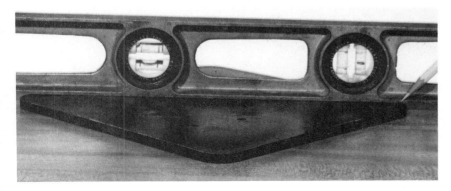

Illus. 21-10. This high-density moulded-plastic mounting plate for a router table is not flat. Check all such mounting plates diagonally, as shown.

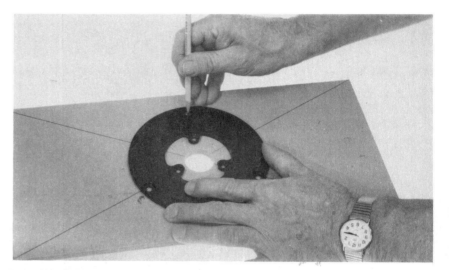

Illus. 21-11. Use your existing sub-base as a pattern for the screw holes.

Illus. 21-12. You may need to purchase longer screws. Do not countersink or counterbore too deeply.

ment, or whatever feature you prefer will be facing the operator's side of the table.

Give some thought, too, to whether you want to locate the router in the center of your table or position it off to one side or end. If you offset it, you then have a choice of space sizes for in-feeding and out-feeding, depending on which side or end of the table you feed from.

Mitre-Gauge Slot

I like a mitre-gauge slot machined into the top that accepts the mitre gauge from my table saw. Some authorities discourage this. I prefer it not so much for the mitre gauge, but for securing and using a special featherboard (Illus. 21-13) and special fixtures and jigs. In fact, on my Ultimate Router Table I cut two mitre-gauge slots, one on each side of the bit opening.

Fences

About 90 percent of the time I use disposable fences. Usually, I just use a short length of straight stock (Illus. 21-13 and 21-14). Often, it's just a 2 × 4 that I joint on one edge and face. This is the most basic type of fence. For grooving and dado-cutting all you really need is a straightedge clamped to the table.

A solid piece of wood used as a straight-line fence needs to be checked frequently for straightness and re-jointed if it becomes distorted. I cut a custom rough-size bit opening and, if I don't like the fence, discard it for another.

Illus. 21-15 shows a fence that is slightly more elaborate than a simple 2 × 4 fence. It has a rabbeted bottom edge for fastening a flat plywood clamping area. A simple

Illus. 21-13. A single straight board makes a serviceable fence. Note the use of the featherboard that anchors itself in a standard ¾" mitre-gauge slot.

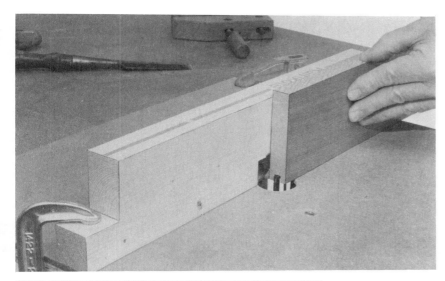

Illus. 21-14. A jointed, straight 2 × 4 clamped flat or on edge (as shown) is the most basic fence.

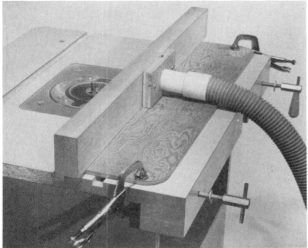

Illus. 21-15. A two-piece fence with vacuum-hose connection. Note the use of quick-action clamps.

block with a hole screwed over the opening for a vacuum-hose connection catches a lot of chips. You can also rejoint the face of this fence when you want to ensure that the cut will be straight.

Illus. 21-16 and 21-17 show a fence made primarily for stopped dadoing and grooving. It has adjustable stops that can be clamped at any point along T-slots cut into the face of the fence. The slots can be used to hold and position other accessories such as guards and hold-downs.

Straight-line edge forming, rabbeting, and many other jobs require the bit to be just partially exposed

beyond the working surface of the fence. In this case, you have to cut an opening for the bit into the fence so you can partially hide it. If you use bits with the same general diameters, you can make a one-piece fence. Illus. 21-19 shows such a fence with a removable guard. Two slip-fit dowels hold the guard in a choice of slots to accommodate work of various thicknesses.

A two-part adjustable fence can be made to accommodate a wide range of bit diameters. The plan for the fence shown in Illus. 21-20 is adapted from a design by Ed Walker, author of the booklet *Table Mounting Routers*. For obvious safety and performance reasons, there should be a minimal amount of open space or clearance around the bit. A Walker-type fence with movable faces permits quick adjustment when you are using a variety of

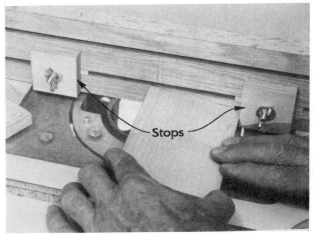

Illus. 21-16. A fence with adjustable stops that slide in T-slots.

Illus. 21-17. End view of the T-slots and the carriage-bolt arrangement for adjusting the stop blocks.

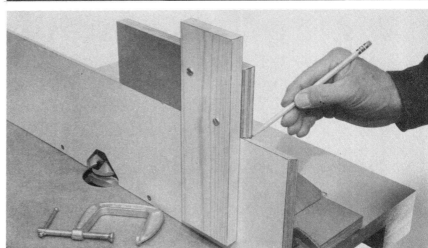

Illus. 21-18. Front view of a typical high fence. Also shown is an optional vertical feeding jig designed for clamping and feeding narrow work such as when tenoning and doing other end-cutting jobs. Note that the jig matches a bevel cut along the top edge of the fence; this improves overall control of the jig and workpiece.

Illus. 21-19. A fence with an adjustable and removable guard.

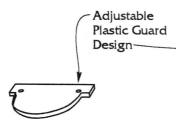

Adjustable
Plastic Guard
Design

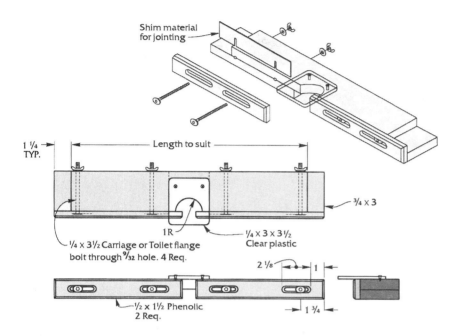

Shim material for jointing

1 1/4 TYP.

Length to suit

3/4 × 3

1R

1/4 × 3 × 3 1/2 Clear plastic

1/4 × 3 1/2 Carriage or Toilet flange bolt through 9/32 hole. 4 Req.

2 1/8

1

1/2 × 1 1/2 Phenolic 2 Req.

1 3/4

Illus. 21-20. A two-part adjustable-fence plan adapted from an original design by Ed Walker.

bit sizes. The out-feed half of the fence can be shimmed out with tagboard or thin plastic for edge-jointing jobs.

High fences (6″ to 10″) are necessary when you are routing stock on edge or in a vertical feed position. Illus. 21-18 and Illus. 21-22 show a shop-made high fence which I use for a variety of common and some unusual routing jobs. The major components are made from flat, ¾″ thick hardwood plywood and are braced with thick screwed-and-glued blocks. The construction is pretty basic, but effective.

You can also devise high fences by simply screwing a wider piece of material to your existing fence. You may want to purchase aluminum castings (Illus. 21-23), adjustment provisions, or a built-in dust-box connection to add to your fence system. Reliable Cutting Tools, San Bernardino, California, or CMT Tools, Tampa, Florida, are two sources of these accessories.

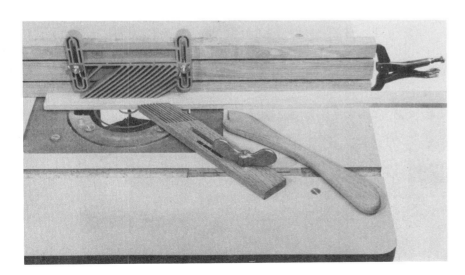

Illus. 21-21. The setup for handling narrow work. The plastic featherboard is from Shopsmith.

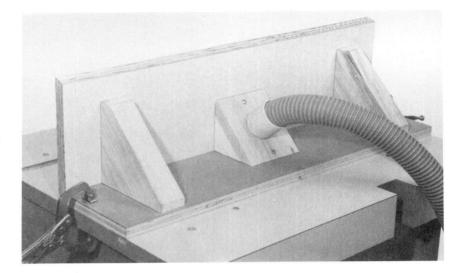

Illus. 21-22. Rear view of the high fence. A dust-and-chip-extraction connection is well worth the effort.

Illus. 21-23. Aluminum castings for router fences similar to this can be purchased.

Dust-Extraction Systems

Dust-extraction provisions are always helpful, but some routing jobs make catching chips difficult—particularly those jobs such as edge-forming with piloted bits and other setups done without a fence and its vacuum connection (Illus. 21-24).

Fulcrums

A fulcrum (also called a starting pin or pivot point) near the bit is an especially helpful item. It is used for the initial feeding of stock into a piloted bit. You pivot the work against the fulcrum as you make the initial cut; this supports and controls the workpiece before it engages with the pilot. Without a fulcrum, you risk damaging the bit by making a jolting cut, as well as causing severe tear-

Illus. 21-24. A dust-collection system used with the short back-and-forth feeding of my box and dovetail-cutting jig. Note the table slots on each side of the bit that guide this jig.

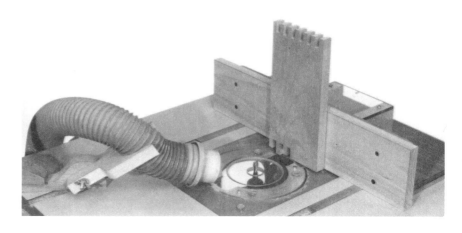

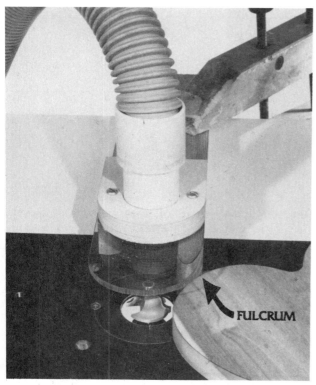

Illus. 21-25. Using the fulcrum to start edge-forming with a piloted bit. Note the clear-plastic guard and vacuum-connection provisions.

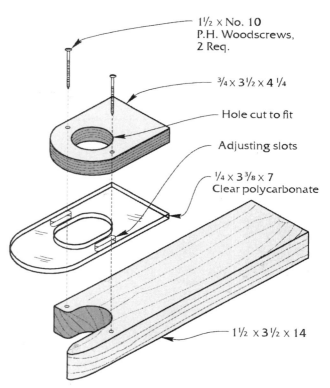

1½ × No. 10 P.H. Woodscrews, 2 Req.

¾ × 3½ × 4¼

Hole cut to fit

Adjusting slots

¼ × 3⅜ × 7 Clear polycarbonate

1½ × 3½ × 14

Illus. 21-27. An exploded view of the components of this jig: a fulcrum, guard, and vacuum-connecting block.

out and kickback. Illus. 21-25–21-27 show a fairly simple but effective combination fulcrum-adjustable-guard-and vacuum-hose arrangement for edge-forming curved, irregular-profiled workpieces.

Auxiliary Switches and Variable-Speed Controls

Auxiliary electrical switches and a variable-speed control can be mounted to a leg, cabinet base, or directly to the underside edge of the top (Illus. 21-28). Make sure that everything is located within convenient reach, but still well protected from accidental start-up. Also consider using a foot switch. It will allow you to keep both hands on the workpiece and eliminate the need to reach with one hand. Be sure to match all switches and the speed control to the amperage rating of your router.

Heavy-Duty Router Table

Heavy-duty routers of at least 3 hp and which weigh up to 17 pounds require well-constructed router tables. I sug-

Illus. 21-26. Normal routing continues after the workpiece engages the pilot of the bit. Note the European-style safety bit, which also minimizes kickback.

Illus. 21-28. You can easily add switches and a variable-speed control to your router table.

Illus. 21-29. A sturdy router table corner mounted and flush to a large workbench. If more fence-to-bit capacity is needed, clamp a longer straightedge directly to the adjoining workbench.

gest that the tops of these tables be a minimum 1″ thick and that they sit on a very substantial support or base. Glue two pieces of ½″ thick panel face-to-face if you cannot purchase 1″ thick material in your area.

A router table setup I made 20 years ago and that I still use for heavy-routing of large timbers, etc., is shown in Illus. 21-29. It is fastened at a corner of my 4′ × 8′ assembly workbench, which has a flat and true work surface. It is made to match the thickness of my assembly workbench. The laminated construction consists of two layers of ¾″ thick particleboard sandwiched by two pieces of ¼″ thick tempered hardboard, for a balanced buildup. The router-table top assembly jets out from my workbench, and is supported on two heavy-steel angles anchored under the workbench with lag bolts. The top of the router table is flush with the workbench surface.

I purchased a second base for my fixed-base router and mounted it permanently under the router table. In fact, I purchased just the base—no handles, knobs, or

sub-bases. Then, using the router itself, I very carefully and accurately inlaid a prepared square of ¼″ thick tempered aluminum. This was set in exactly flush and secured with flathead machine screws through the laminated table. These flathead screws were fit directly into the holes in the base which are normally used to attach a sub-base, thus attaching the square to the base. All exposed wood and edges were sealed, varnished, and waxed.

My router table is virtually warp-proof and moves very little with changes in humidity. Remember that the router table fence does not have to be parallel to the table edge. Consequently, if more distance between the fence and bit is required, I can clamp a longer straightedge obliquely across the assembly bench. Thus, I can utilize the full 4′ × 8′ bench for in-feed or out-feed support when necessary. This arrangement is particularly advantageous when it is necessary to support the leading end of long, heavy workpieces.

It is almost impossible to make freestanding folding or portable router tables that are really suitable for big routers. It may be best to design a top that can be clamped or supported from or on a heavy workbench, as just described. Another alternative is to design a top as an extension wing for your stationary table saw.

For big plunge routers, remember to take advantage of a fine-depth-of-cut adjustment accessory if one is not standard on your router (Illus. 21-30). This makes adjustments much easier than lifting the router out of its "cradle" every time you want to change the depth of cut.

Illus. 21-30. The depth-of-cut control is standard on some routers and available as an accessory for most others.

22
Router-Table Safety Techniques

Whenever setting up and operating a router table, approach *every* aspect of use with an awareness of the proper safety procedures. Always use whatever safety devices are appropriate for the job at hand. Illus. 22-1 shows some items that you can make or purchase.

Chapter 8 describes safety techniques for using a hand-held router. These techniques also apply to router-table use. This chapter concentrates on safety techniques for five areas of concern when using the router table: (1) feed direction; (2) cutting small and narrow stock; (3) making stopped and deep blind cuts that require lowering the stock onto a rotating cutter; (4) vertical (on edge) routing; and (5) using large-diameter bits.

Feed Direction

Use a well-clamped fence for all straight-line work, even when using piloted bits. Plan to make the cut so that the direction of feed is *against* the rotation of the bit (Illus. 22-2). The force or thrust from the bit's rotation should pull the workpiece against the fence, not push it away, which is extremely dangerous and happens when you feed with the rotation of the bit. The latter situation usually happens when you trap the work between the fence and the bit (Illus. 22-3).

Some router users try to remember the correct feed direction by routing from right to left against the bit's counterclockwise rotation. Others rout to the left side of the bit. If in doubt, scratch a curved arrow into your router plate showing the bit's rotation.

When making multiple passes to widen grooves, mortises, dadoes, etc., be certain that you do not inadvertently trap a portion of the workpiece. In such situations, you must be sure that the bit is working on the *safe* side of the initial cut so the cutting and thrust action pull the work towards the fence, not push it away (Illus. 22-4). If

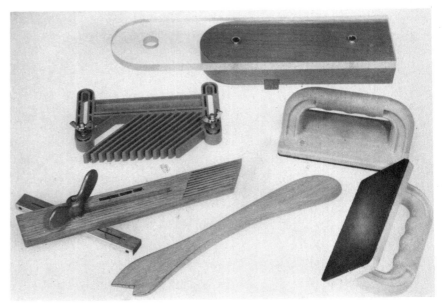

Illus. 22-1. A few important router-table safety devices. Top: a guard. Left: featherboards. Bottom: a push stick. Right: push blocks.

Illus. 22-2. A correct, safe setup for straight-line edge-routing with a fence. The force or thrust of the rotating bit pulls the workpiece against the fence.

Illus. 22-3. This is a dangerous setup! It traps the workpiece between the fence and the bit. This creates an uphill-cutting action, an uncontrollable feeding situation where the bit will grab the work and throw it in any direction with great speed and force.

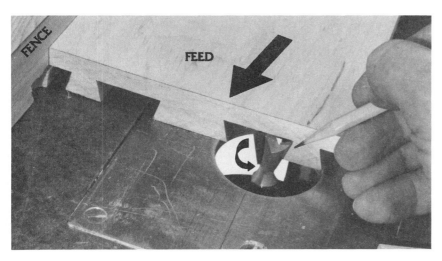

Illus. 22-4. The safe, correct setup for widening cuts with multiple passes. The thrust pulls the workpiece towards the fence.

you cut on the wrong side of the groove, you are again uphill-cutting and have a potentially dangerous situation (Illus. 22-5).

Illus. 22-5. **DANGER! This is a dangerous setup for widening grooves, etc., with multiple passes. The bit will grab the trapped workpiece and propel it in any direction.**

Edge-Forming Curved Profiles

Take the same precautions concerning a feed direction when edge-forming curved profiles, to ensure that you feed *against* bit rotation and not with it. Illus. 22-6 and 22-7 show the correct feed directions for forming outside edges and edge-forming around inside openings.

Straight-Line Routing

Straight-line routing with piloted bits should generally still be done with the fence (Illus. 22-8). When you are starting the cut, the fence allows you to control the workpiece before the leading edge of the stock reaches the bearing. Routing with nonpiloted bits requires a fence or other means to control the cutting action. Illus. 22-8–22-12 show a variety of setups for safe routing of different jobs.

Starting Pins and Fulcrums

Starting pins and fulcrums are necessary to control the stock until it is advanced to the point where it reaches the

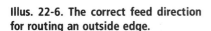

Illus. 22-6. The correct feed direction for routing an outside edge.

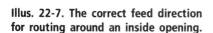

Illus. 22-7. The correct feed direction for routing around an inside opening.

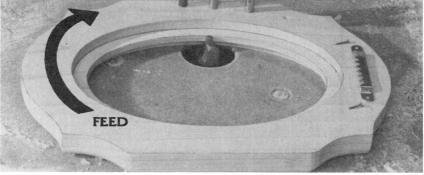

Illus. 22-8. For most straight-line work with piloted bits, I still prefer to use the fence to guide the work. The fence is adjusted to the bit.

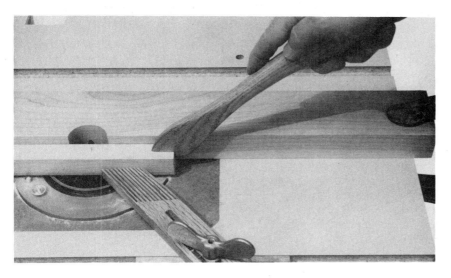

Illus. 22-9. Safe routing of narrow stock. Make sure that the push stick does not touch the bit.

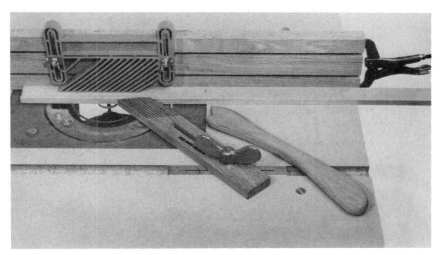

Illus. 22-10. A safe setup for routing narrow and thin stock. Make sure that the featherboards or the push stick do not touch the bit.

Illus. 22-11 (left). Vertical routing of narrow stock sometimes is necessary, but it should only be done with a sliding fixture and a high fence, as shown here. Illus. 22-12 (above). When making deep dovetail grooves and dadoes, first remove their excess waste with a straight bit before using the dovetail bit. Because of their design, dovetail bits are weaker than straight-cutting bits. Removing the excess waste first will prevent the bit from getting trapped in a chip-impacted cut.

Illus. 22-13. Maintain pressure against the fulcrum while advancing the work-piece into the rotating bit.

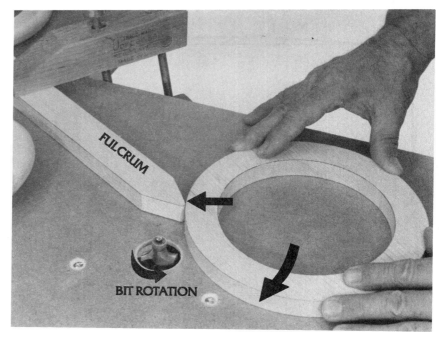

bearing or pilot. Clamp a pointed board to the table (Illus. 22-13) or set steel starting pins into the router plate whenever possible (Illus. 22-14 and 22-15).

Cutting Small Pieces

Small pieces must be handled with extreme caution. If they are too small to hold safely, don't rout them or be sure to use safe fixtures (Illus. 22-11). Don't ever position your hands close to the bit.

Shop Aids

Push Blocks

Push blocks with rubber friction pads on their undersides (Illus. 22-16) can be used as a hold-down and feeding tool for *some* small pieces and when making

shallow cuts. A backer push block (Illus. 22-17) has two important functions. First, it eliminates tear-out or splintering as the trailing edge of the workpiece passes the bit when crosscutting types of cuts are made. Second, it guides narrow stock safely past the router bit perpendicular to the fence, to ensure square cuts.

Spacers, Auxiliary Tops, and Jigs

Using pieces of hardboard or plywood panel spacers can often eliminate the need to change setups of bit depths or fence locations when making multiple passes. These items are especially useful when you are using big bits or when you need to make certain deep cuts. You can also use an auxiliary fence or tabletop surface to reduce the clearance opening around the bit (Illus. 22-20 and 22-21).

Illus. 22-14. Safety apparatus necessary for all edge-forming jobs include the plastic guard positioned directly over the bit and a fulcrum or starting pin.

Illus. 22-15. Here the operator is starting a crown cut on a raised panel. The guard is in place, the work is held tightly against the fulcrum starting pin, the bit height is set for a shallow cut, and a slow router speed is being used.

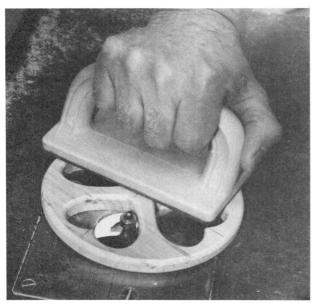

Illus. 22-16. Use these push blocks with rubber friction pads on their bottoms to hold and feed small parts and to keep your fingers well away from the bit.

Illus. 22-18. Splintering on the exit side of this vertical routing cut.

HIGH FENCE

Illus. 22-17. A shop-made backer push block not only guides narrow work past the bit as shown, it eliminates tear-out or splintering on the exit side of the cut. When the block gets "chewed" up, mount the handle to another square block.

Illus. 22-19. A scrap backer supports the workpiece to reduce splintering and tear-out.

BACKER

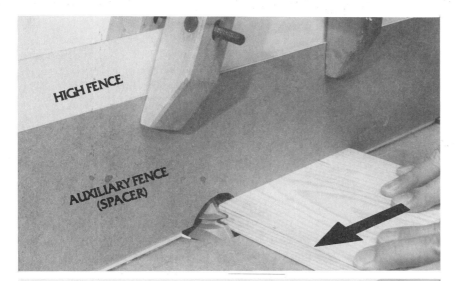

Illus. 22-20. This big lock mitre bit removes a lot of material. To make the cut in two passes with just one fence setting, use an auxiliary fence (spacer) clamped as shown. Remove it after the first pass.

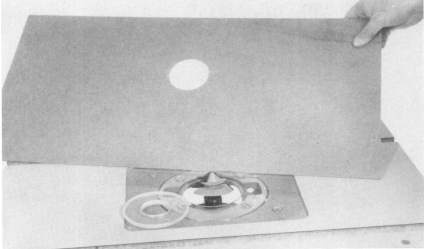

Illus. 22-21. This auxiliary table when clamped or set down with double-faced tape will be used to reduce the clearance hole around the bit.

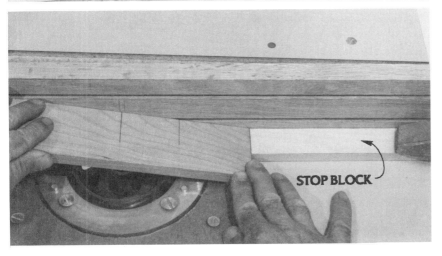

Illus 22-22. Use a stop block when making any cut where the workpiece must be lowered into a rotating bit.

Stop Blocks

When mortising, slotting, and making stopped dadoes, rabbets, and grooves with a router table, you must often lower the workpiece into a rotating bit. A stop block appropriately placed and clamped behind the work (Illus. 22-22) will prove helpful.

Setup Blocks

Setup blocks save time. Next time you have made a cut, make a setup block afterwards. This way, the next time you want to make the same cut, simply adjust the fence and bit height to the setup block (Illus. 22-23).

Illus. 22-23. Use a setup block to adjust the fence and bit depth.

23
Router Table Operations

This chapter explores many operations and a wide range of different cuts that can routinely be performed on conventional router tables. To clearly show the cuts, most of the photographs in this chapter show router tables without guards, hold-downs, or other accessories that are normally recommended for operator safety. You, obviously, should use these accessories. Biscuit joinery, dovetailing, box joints, vacuum-template duplicating work, and other joinery involving specialized techniques are covered in future chapters.

Basic Techniques

The following sections do not cover every possible cut or joint, but should provide a good overview of the techniques that can be applied to your own table-routing jobs.

Fence Work

Fence work involves all straight-line cuts. Operations include those in which the bit is only partially exposed beyond the working face of the fence and those jobs where the fence is set some distance from a totally exposed bit. Illus. 23-1–23-4 show how a tongue, rabbet, and a sliding dovetail cut are made. Illus. 23-5 shows an interesting bead-and-rabbet corner joint.

Cutting Grooves and Slots

Grooves and slots are all cut in generally the same way. The depth of cut and fence location (Illus. 23-6 and 23-7) must be carefully set. Measure as accurately as possible and then make a test cut.

Illus. 23-1. Cutting the tongue for a tongue-and-groove joint.

Illus. 23-2. Making a dovetail cut for a dado-dovetail joint.

Illus. 23-3. This one-piece fence setup shows a ¾" bit with zero fence clearance for routing a shallow rabbet equal to the bit width.

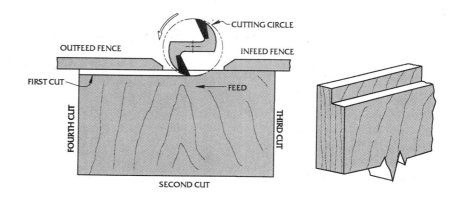

Illus. 23-4. A straight-line fence setup for edge-forming when using a split-type shaper fence with one or both halves adjustable. When edge-forming all four edges of a board, do the end grains first.

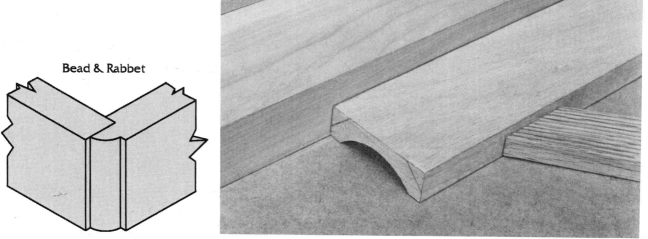

Bead & Rabbet

Illus. 23-5 (above left). Here's an interesting design for a corner joint. Illus. 23-6 (above right). Making a cove cut. Note the straight board fence and featherboard hold-in.

Illus. 23-7. Carefully adjust the fence to the cutting edge of the bit. Double-check all critical adjustments with a trial cut.

Stopped Cuts

To set up for stopped cuts of any kind, draw a registration mark or marks on the fence (Illus. 23-8). These marks are used to indicate where cuts are to begin or to stop. Your feed can stop at just a line (Illus. 23-9) or be set up so the work strikes a preset stop block clamped to the table or fence (Illus. 23-10 and 23-11). Using a stop block to locate the start of a totally blind or stopped cut is a recommended safety precaution, and is described in Chapter 22. You must keep the workpiece against the fence and stop while simultaneously lowering it onto a rotating bit (Illus. 23-12). This is obviously a practice you should avoid with small-size stock. Illus. 23-13 and 23-14 show an open- or through-slot cutting operation.

Illus. 23-8. Project two registration marks on the fence equal to the cutting diameter of the bit.

Illus. 23-9. Making a shallow groove for shelf hardware. Note the three marks on the board fence visible at the lower left of the photo. Two indicate the bit diameter, and the one at the far left indicates the length of the cut.

Illus. 23-10. Here a stop block is clamped to the fence to stop the cut. A stop can similarly be located to support work for starting the cut. Stops are recommended for production jobs.

Illus. 23-11. Completed cuts of open mortises or stopped grooves.

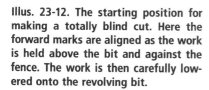 Illus. 23-12. The starting position for making a totally blind cut. Here the forward marks are aligned as the work is held above the bit and against the fence. The work is then carefully lowered onto the revolving bit.

Illus. 23-13. Once the workpiece has been lowered over the bit, the work is advanced as in normal routing. Note the stop or starting block at the right.

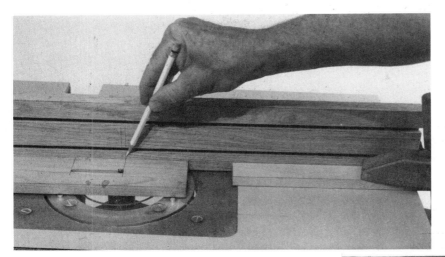

Illus. 23-14. Stop feeding the workpiece when the registration marks indicating the length of the slot line up with the near-side bit mark on the fence.

Cutting Mortises

Cutting mortises (Illus. 23-15) on the router table essentially involves similar blind routing procedures. Use registration lines on the fence to make the setup and to locate two stop blocks for making the cut. Because you must lower the workpiece into a rotating bit, deep mortising is not recommended. A hand-held plunge router is generally safer and easier to use. However, my router table described in Chapter 24, which incorporates an up-plunge router, is an ideal table for deep mortising (Illus. 23-15–23-17) and for production mortising.

Illus. 23-16. The setup for mortising with an upcut router. Registration marks indicate the length of the mortise. Note the use of the stop and featherboard.

Illus. 23-15. A ¾" deep mortise that was cut on a router table in which an upcut plunge router was used. This table is described in Chapter 24.

Illus. 23-17. The stop controls the length of the mortise.

Tenoning

Cutting tenons is very easy with a good router table. You can use the actual mortise to establish the bit height for the tenon shoulder and cheek cuts (Illus. 23-18–23-21).

Edge-Jointing

Edge-jointing on the router table can be done in two basic ways. You can either use a straight or spiral bit and a special fence with an offset out-feed face (Illus. 23-22 and Illus. 23-23), or a flush trim bit and a straightedge (Illus. 23-24).

Illus. 23-18. Set the bit height for cutting the tenons to a previously mortised part, as shown.

Illus. 23-19. Use a right-angle pusher block to eliminate splintering on the exit side and to keep the workpiece square to the fence. Note the guard.

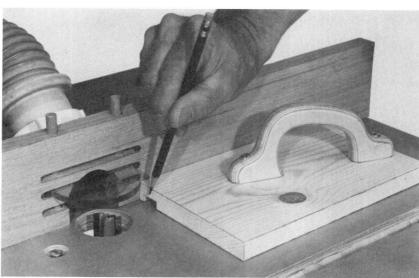

Illus. 23-20. The cut made in the pusher block. When "chewed-up" from being used for other cuts, the pusher block may not provide backing where wanted. At this point, saw some backing off the front edge or cut a new block and fasten the handle to it.

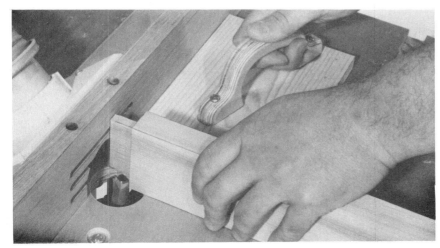

Illus. 23-21. Cutting the tenon to width.

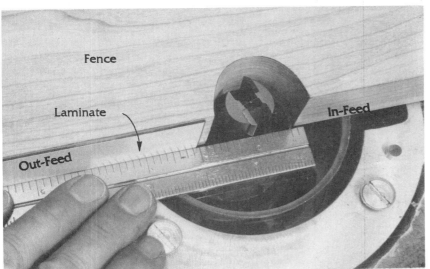

Illus. 23-22. This easy-to-make jointing fence is simply a one-piece board with plastic laminate on the out-feed half. Adjust the fence so the out-feed surface is tangent to the cutting circle of the bit.

Illus. 23-23. Jointing an edge. The best results are obtained with larger-diameter bits. Notice the use of a featherboard.

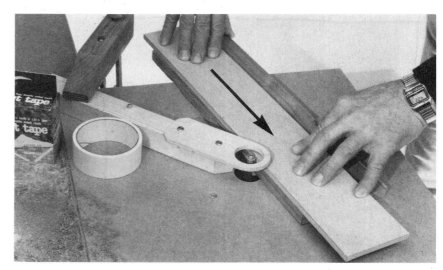

Illus. 23-24. Jointing with a ball-bearing-guided flush-trimming bit. Secure a straightedge to the top of your workpiece with double-faced tape. Note the guard and feed direction.

Full-Edge Shaping When you are shaping full edges (Illus. 23-25), as when jointing, you have to use the out-feed half of the fence to support and guide the surface of the workpiece cut away by the bit. Use either an adjustable split fence or extend the out-feed side of the fence you have for the appropriate distance. However, in certain jobs you can compromise. If you don't cut away all of the edge, you don't need an offset out-feed fence (Illus. 23-26). Employing the same idea allows you to make your own dowels with a round-over bit (Illus. 23-27).

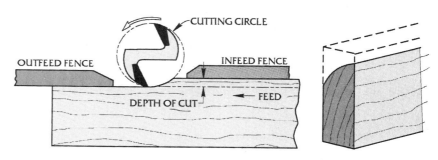

Illus. 23-25. Full-edge shaping. The out-feed end of the fence must be offset to support the cut surface.

Illus. 23-26. Just a tiny flat is left uncut on this fully shaped moulding so that surface will support the work on the out-feed side of the cutter. Note the use of featherboards.

Illus. 23-27. Nearly perfectly round dowels (usually as round as those you can buy) are made in four passes with this basic setup and a round-over bit.

Squaring and End-Trimming

Squaring and end-trimming with a good bit can produce spectacular cuts. This work can be performed with the aid of a sliding fixture. The mitre-gauge-like fixture shown in Illus. 23-28 is simply a clamping block secured to a thin plywood base with a hardwood strip under it that follows in a slot cut into the table. A similar fixture shown in Illus. 23-29 and 23-30 is designed for a production job requiring many pieces of identical lengths with crisply cut end grains.

Illus. 23-28. End-trimming and squaring with a shop-made fixture that functions like a mitre gauge.

Illus. 23-29. A similar fixture, but this one is made for squaring rough-sawn pieces to a specific length. It has a thin wood spring stop which compresses out of the way while the first end is being cut.

Illus. 23-30. With one end cut, flip the work so the squared end butts against the wood spring stop for making the second cut. Every piece will be exactly the same length.

Pattern- and Template-Routing

Pattern- and template-routing to reproduce parts can be done very conveniently on the router table. Use essentially the same techniques as employed with hand-held work discussed in Chapter 13. Remember to feed the work counterclockwise into the bit when routing around the outside perimeter (Illus. 23-31). Feed the work into the bit in a clockwise direction when trimming inside cuts (Illus. 23-32 and 23-33).

Illus. 23-31. A quarter pattern nailed to the back of a picture-frame workpiece ensures identical cuts all around the workpiece. To cut the outside edge, feed the work counterclockwise into the bit.

Illus. 23-32. Cleaning up the rough-cut inside openings of a multiple picture-frame project. The work is fed clockwise around the bit to make these inside trimming cuts.

Illus. 23-33. A close-up shows the clean, smooth cuts made with the trimming bit.

Cutting Round Tenons and Shaping Dowels

Cutting round tenons on the ends of dowels and making decorative cuts in round stock are both novel and practical table-routing jobs. Many of the cuts shown in Illus. 23-34 and 23-35 were made with the aid of a simple V-block setup (Illus. 23-36). This fixture consists of a V-block (with a bored hole for the bit) fastened with a screw or glue to a suitable sheet of ⅛" thick hardboard. This fixture is clamped to the router table. An independent fence is set, clamped, and used as a stop to limit the length of the tenon or to locate other cuts.

Piloted bits and ball-bearing-guided bits can be used to cut decorative dowel ends (Illus. 23-37). In this situation, a V-block does not have to be used. A simple support block or fence works just as effectively. Feed the dowel into the bit and then rotate it against the rotation. Some typical dowel end cuts made with various piloted bits are shown in Illus. 23-38.

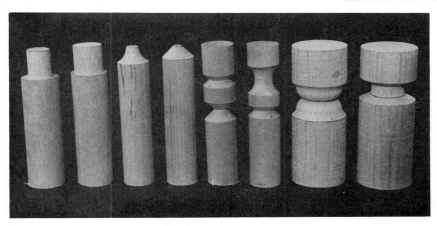

Illus. 23-34. These dowel ends were machined on a router table. Round tenons are shown on the left. The other decorative cuts have good applications for wood toys and chess sets.

Illus. 23-35. More decorative shapes cut into dowel stock, and the bits used.

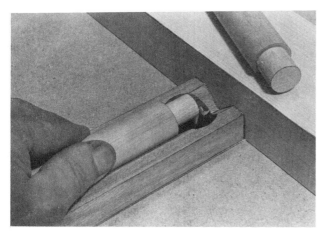

Illus. 23-36. A close-up look at the setup for cutting round tenons with a straight-cutting bit. Note how the fence acts as a stop.

Support Fence

Illus. 23-37. Decorative end cuts as well as tenons can be made by rotating the dowel against a support fence positioned as shown.

Illus. 23-38. Some interesting dowel end cuts made with typical piloted edge-forming bits.

Special Cabinet Joinery

This class of work involves setting up and using some of the more popular special bits such as bits for cutting drawer lock joints, lock mitre joints, cope stiles and rails, and raised panels.

Lock Mitre Joints and Drawer Lock Joints

Both component parts of lock mitre joints (Illus. 23-39) and drawer lock joints (Illus. 23-40) are cut with just one bit. Trial-and-error setups with some test cuts on waste material are usually necessary. These cuts are similar in that one part of the joint is cut with the stock fed flat on the table, and that the second component of the joint is made by passing the stock vertically or on its edge over the same bit. In both instances, a true flat table (especially in the mounting-plate area) and a perfectly perpendicular and straight fence are essential to good results.

Illus. 23-41 and 23-42 show how to make the cuts for a lock mitre joint. This bit has a large cutting diameter, and making the full cuts in just one pass may not be wise. I suggest clamping a ¼″ thick auxiliary fence to the existing fence for making the first pass on both parts. Remove it for making the second and final pass. Using this shim or spacer system eliminates the need to reset the fence for

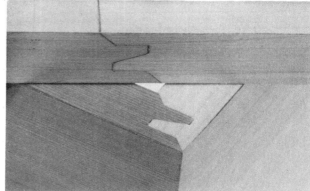

Illus. 23-39. Lock-mitre-and-glue joint. All cuts were made with just one bit, shown on top.

Illus. 23-40. Both components of these joints for making flush (above) or lipped (below) drawers were cut with one bit, a drawer-lock-joint cutter. The one shown here has a European safety design and is available from CMT Tools in Tampa, Florida.

Illus. 23-41. The horizontal routing of one half of a lock mitre joint. When making this cut, make sure the workpiece's inside face is down and that the cut is centered in the stock.

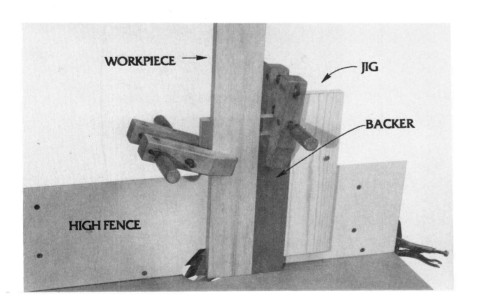

Illus. 23-42. Make the second component of a lock mitre joint by cutting vertically with the workpiece's inside face against the fence. The backer eliminates tear-out or splintering as the trailing edge passes the bit.

making new cuts at an increased depth. Be sure to always use support backers to eliminate tear-out and splintering of the trailing edge as the work exits past the bit. Refer to pages 264 and 265 for more information about machining lock-mitre joints.

Illus. 23-43 and 23-44 show the same two kinds of cuts involved in making either flush or lipped drawers using the drawer lock-joint bit.

Coped Rail-and-Stile Joints

Coped rail-and-stile joints (Illus. 23-47) used to be made exclusively on a shaper. Today, bit sets and the router table make it relatively easy to cut these joints (Illus. 23-48). Remember, stiles are the vertical members of a frame, and rails are the horizontal members. Two

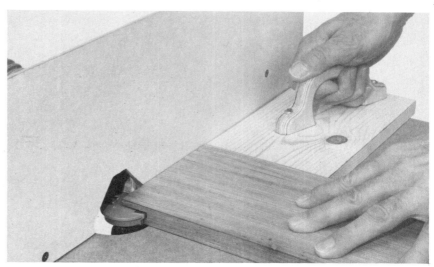

Illus. 23-43. The drawer front is routed facedown. Use a right-angle push block to eliminate tear-out and for feeding narrow workpieces.

Illus. 23-44. Routing the drawer sides vertically with a high fence. Use the supporting guide to feed drawer sides of narrow widths.

Illus. 23-45. Two joint-making bits from Sears with ¼″ diameter shanks. At left is a drawer-joint bit. At right is a bit for cutting tongues and grooves.

Illus. 23-46. Dovetail rabbet cuts make a good, quickly nailed drawer corner joint for either flush (shown) or lipped drawer fronts.

Illus. 23-47. An assembled door with a raised panel and a coped stile-and-rail frame. Note the shaped inside edges of the frame components.

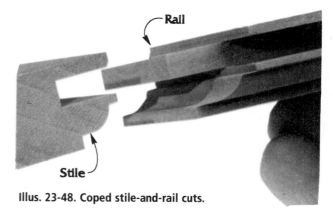

Illus. 23-48. Coped stile-and-rail cuts.

matching (male and female) cutting bits (Illus. 23-49) are sold in sets to make all the necessary cuts. Also available for this work is a single bit, called a "reversible" bit, which also makes the same coped (matching-shape) cuts. However, you must reverse the cutters on the arbor, and usually you need to insert shims on a trial-and-error basis to get a perfect, matched cut.

Illus. 23-49. Factory-matched bit sets for cutting coped rails and stiles are far easier to use than a single "reversible" bit.

When using the matched two-bit set, you must make just two simple setups with each bit. The first involves the appropriate height. This is easy if you are using the bit sets shown in Illus. 23-49 and 23-50, because they are profiled for only ¾″ to ⅞″ thick stock. The second setup involves adjusting the fence so its working face is perfectly flush with or tangent to the bearings (Illus. 23-51).

All cuts are made with the face sides of the frame stock placed down. For new users, it is a good idea to make a few practice cuts with each bit to understand how they function together, because it can be confusing the first time around. I like to cut the coped ends on the rails

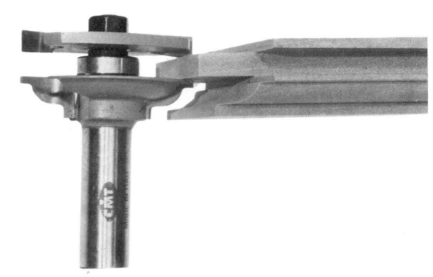

Illus. 23-50. Only the ends of the rails get cut with the rail bit.

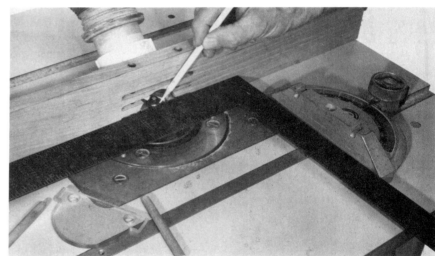

Illus. 23-51. The bearings of the bits should be flush to the face of the fence. Also, make sure the fence is square to a mitre gauge or use a pusher block. Note the guard components at lower left.

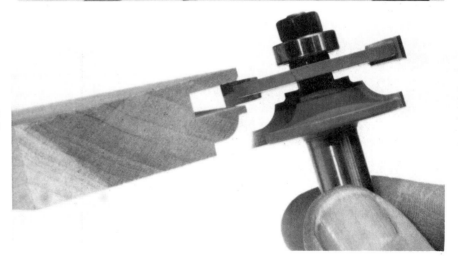

Illus. 23-52. The inside edges of both the stile and rail are cut with the stile bit.

first. However, the easiest way to set up the bit is to adjust it to its appropriate height using a previously stile-cut test piece (Illus. 23-52 and 23-53).

Make the tenon and coped cuts on the ends of the horizontal rail pieces only. These are the only cuts the rail bit is used for (Illus. 23-54). Finally, rout the inside edges of the stiles (Illus. 23-55) and, with the same bit, rout the inside edges of the coped-cut rails.

Illus. 23-53. After installing the rail-cutting bit, use a setup block or a previously cut stile, as shown here, to set the height of the bit. Align the groove to the corners of the cutters that make the tenon or tongue cut.

Illus. 23-54. Make the coped cuts on both ends of all rail pieces with their face sides down. Note the use of a scrap backer to eliminate tear-out as the work exits past the bit.

Illus. 23-55. Routing a stile with its face down. Note the guard and featherboard hold-in.

Raised Panels

Raised panels (Illus. 23-56) typically complement coped stile-and-rail frames to make beautiful and stylish cabinetry doors and architectural panels. There are two types of panel-raising bits: vertical (Illus. 23-57) and horizontal. Both are designed to remove a lot of material, but it's best to make successive passes at shallower depths. Illus. 23-58–23-60 show the setup and technique employed to make raised panels using the vertical cutters to machine panels as they are fed on their edge past the bit.

Horizontal raised-panel routing (Illus. 23-61) requires much caution because these bits have the largest cutting diameters of all bits used with the router. It is essential that you use sound stock, fully controlled, shallow cuts, and slow router speeds. Review all safety instructions and the manufacturer's warnings and directions before using these bits. Make straight-line cuts with the aid of the fence. Even though these bits have pilots, hide most of the bit in the fence whenever you have the opportunity.

The major advantage of the horizontal-cutting bit, however, is that the pilot does allow you to make cuts with the bearing following an irregular edge. This feature permits you to make arched raised panels (Illus. 23-62). Here all possible safety precautions should be employed because if a fence is not used, the entire bit is standing alone in the middle of the table. Use a guard, a starting pin, and the correct feed direction (Illus. 23-63).

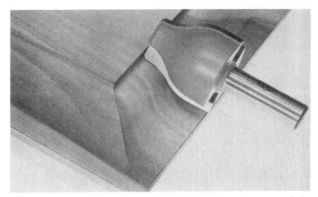

Illus. 23-56. A vertical raised-panel bit and a close-up of the smooth, finished cuts produced with and across the grain.

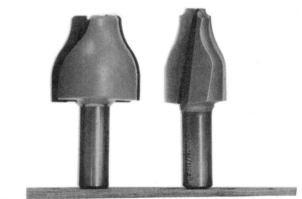

Illus. 23-57. Vertical raised-panel bits are available in various cutting profiles.

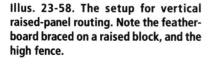

Illus. 23-58. The setup for vertical raised-panel routing. Note the featherboard braced on a raised block, and the high fence.

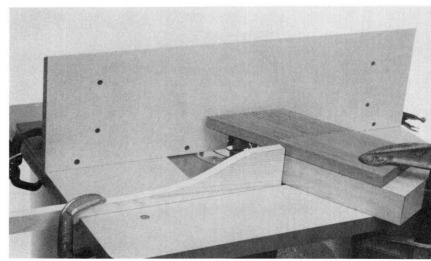

Illus. 23-59. A close-up look at the setup. Here the bit and fence are adjusted for a shallow cut for the final pass.

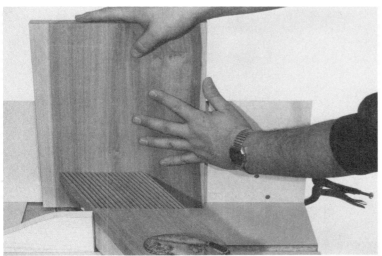

Illus. 23-60. Always rout the end grains first.

Illus. 23-61. Panel-raising with a horizontal panel-raising bit. Be extremely careful when performing this operation. Make multiple passes at shallow depths and use slow router speeds.

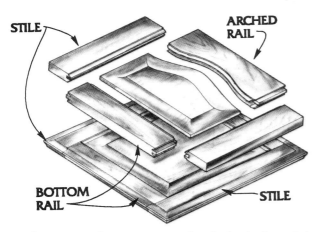

Illus. 23-62. The components of arched raised-panel doors with cope-joint rail-and-stile frames. (Drawing courtesy of Unique Machine and Tool Co.)

Illus. 23-63. The setup for routing arched raised panels with a horizontal-cutting bit. The face of the panel is placed down. Make successive passes at shallow depths of cut until you complete the full cut. Note the use of a starting pin. The guard was removed for photo clarity. You should use a guard.

24
Router Tables for Special Operations

Sometimes, the need or desire to make unusual or special cuts with your router may dictate the general design of the router table. Conventional router tables are configured so the router is held vertically and upside down. However, there are router tables that have been designed so the router can be used horizontally, obliquely, or angularly. This chapter looks at *Woodsmith®* magazine's popular mortising table, which incorporates an adjustable horizontal router. I also describe and illustrate two other router tables designed to place the router in positions that are angular to the worktable, and preview a working prototype of my latest creation, the Up-Plunge Router Table.

Woodsmith Mortising Table

The Woodsmith mortising table (Illus. 24-1 and 24-2) is made from a kit (Illus. 24-3) or plans published in *Woodsmith®* magazine (issue number 67). Both can be obtained from Woodsmith Corporation, Des Moines, Iowa. A hardware kit consisting of the mounting plate, guard, knobs, plans, and instructions is also available.

Although this table cuts mortises easily, it is not limited to just mortising. It's perfect for grooving edges or ends (Illus. 24-4), tenoning (Illus. 24-5), making end lap cuts, and much more. The workpiece can be fed flat across a table surface that's approximately 12″ × 16″, and it also has a mitre-gauge slot. With the guard removed, the work can also be fed vertically against the upright panel.

One problem with this design is that over one inch in

Illus. 24-1. The working side of Woodsmith's® bench-top mortising table.

Illus. 24-2. This backside view shows the router attached to an adjustable router base plate.

Illus. 24-3. Woodsmith sells the entire kit, which only requires assembly. You can also make your own mortising table from Woodsmith's plans and/or order just the hardware shown in the lower part of this photograph.

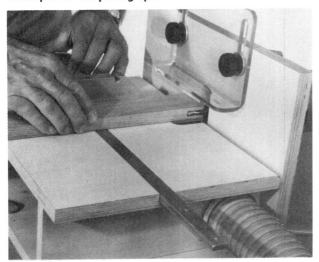

Illus. 24-4. Routing an open mortise.

the router's depth of cut is lost due to the combined thickness of the vertical back panel and the router base plate. You need long mortising bits if you require more

Illus. 24-5. Cutting a tenon on Woodsmith's mortising table.

than average depth of cut. Deep cuts have to be made in multiple, shallower passes, to avoid chattering.

Horizontally Tilting Router Table

The horizontally tilting router table shown in Illus. 24-6 and 24-7 is similar to the Woodsmith® mortising table except this one incorporates the Bryco® cast-aluminum

Illus. 24-6. The "business" side of the angular, horizontal bench-top router table.

Illus. 24-7. This backside view shows the Bryco router bracket fastened to a vertically adjustable pivot plate.

tilting router bracket (Illus. 24-8). It can do all the Woodsmith table can do, plus slightly more (Illus. 24-9). However, it can only be used with fixed-base router motors with metal housings. This router table is not as easy to set up and the depth-of-cut adjustments cannot be made as quickly because the table is being used without the router's own depth-setting adjustment mechanism.

Once set up, this table allows you to get many more cutting configurations from a single bit. A different angle setting cuts a new profile. For example, the straight edge of a roundnose bit can be utilized to chamfer edges (Illus. 24-10) or to make raised-panel-type cuts (Illus. 24-11). Illus. 24-12 gives all of the essential construction details for making this table.

Illus. 24-8. The Bryco cast-aluminum router tilting bracket is available from various mail-order catalogues or directly from Bryco Inc., Champaign, Illinois. This company recommends that the bracket be used mounted to the top surface of a router table.

Illus. 24-9. Straight and angular tenons made with the tilting horizontal router table.

Illus. 24-10. Chamfering with the straight edge of a roundnose bit.

Illus. 24-11. Making raised-panel-type cuts with a roundnose bit.

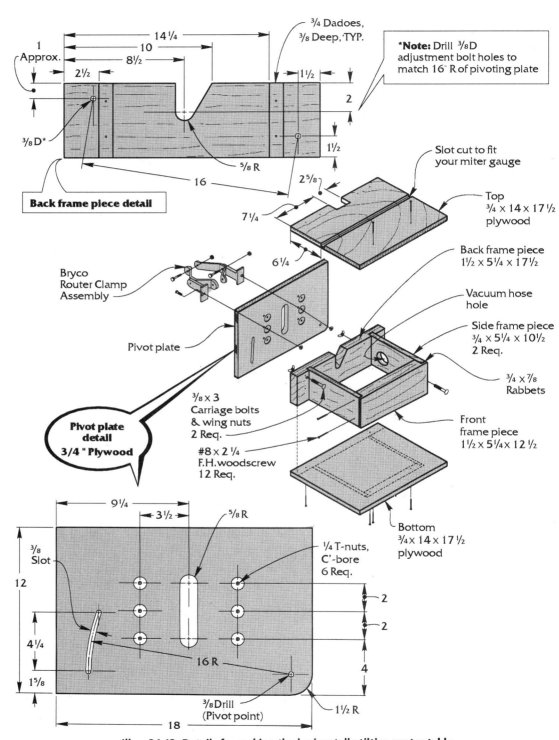

Back frame piece detail

Pivot plate detail
3/4 " Plywood

Illus. 24-12. Details for making the horizontally tilting router table.

Vertically Tilting Router Table

Illus. 24-13–24-16 show a vertically tilting bench-top router table featuring a shop-made router bracket. The router can be centered in the top or near the end of this table. This table was designed primarily to decorate dowels and for making chess pieces, but I've found it's also great for a variety of other unusual routing jobs. Here, too, you can use one bit to cut a variety of different shapes.

Illus. 24-15. A more centralized router position is obtained simply by mounting the router clamping-and-pivoting block to the inside, as shown.

Illus. 24-13. This vertically tilting router table is designed so the router can be centered in the top or near an edge (as shown).

Use an auxiliary top of ⅛″ to ¼″ thick tempered hardboard to reduce the oblong bit openings in the top. Simply attach it with double-faced tape. Fences, guide blocks, or fixtures of any kind can be clamped on top of the auxiliary table. Note the sufficient clamping space provided all around the top shown in Illus. 24-13.

Illus. 24-18 and 24-19 show the table set up to form the ends of dowels with a typical ball-bearing-pilot ogee bit. The router is operating in the conventional vertical position. The router is mounted to the table in the position that is nearest the table edge, to intentionally minimize the in-feed working area.

Illus. 24-20 shows a few examples of many interesting dowel cuts you can make using nonpiloted bits with the router clamped obliquely to the table. The setup for feeding and rotating the dowel is shown in Illus. 24-21 and 24-22.

Illus. 24-14. A close look at the bracket and hose clamp motor-mounting system. Note that the router tilts right and left, but it also has some vertical range.

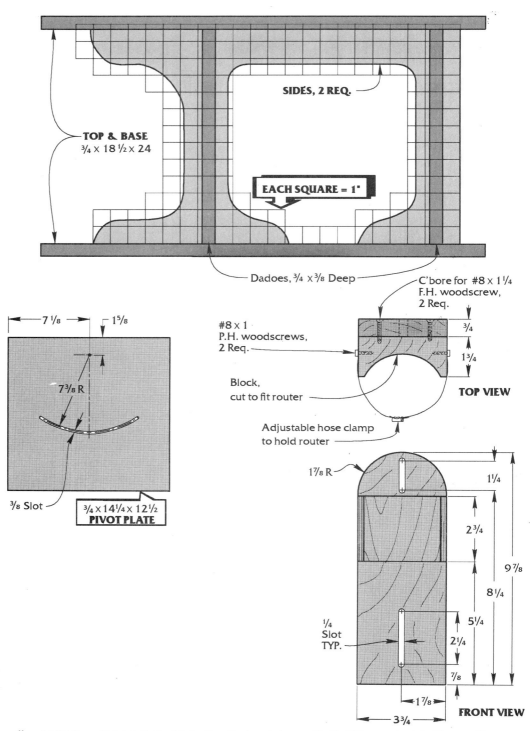

TOP & BASE
3/4 × 18 1/2 × 24

SIDES, 2 REQ.

EACH SQUARE = 1"

Dadoes, 3/4 × 3/8 Deep

7 1/8　　1 5/8

7 3/8 R

3/8 Slot

3/4 × 14 1/4 × 12 1/2
PIVOT PLATE

C'bore for #8 × 1 1/4
F.H. woodscrew,
2 Req.

#8 × 1
P.H. woodscrews,
2 Req.

3/4

1 3/4

Block,
cut to fit router

TOP VIEW

Adjustable hose clamp
to hold router

1 7/8 R

1 1/4

2 3/4

9 7/8

8 1/4

1/4
Slot
TYP.

5 1/4

2 1/4

7/8

1 7/8

3 3/4

FRONT VIEW

Illus. 24-16. Essential details for fabricating the bench-top vertically tilting router tale. The specific top bit-hole location details are left to the discretion of the builder, because routers and the mounting block will vary. The top and bottom measure 18½" × 24".

Illus. 24-17. This piece of thin tempered hardboard used as an auxiliary working surface has been stuck down on the router table's top with double-faced tape to reduce the oblong bit hole in the top.

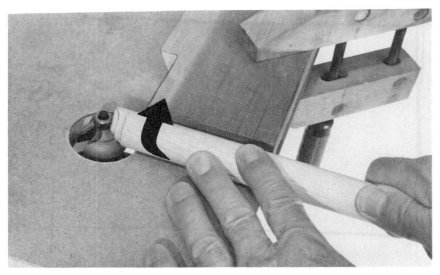

Illus. 24-18. Forming the ends of dowels with a piloted bit and a simple support fence. Rotate the dowel so it is turned downwards and fed against the bit's rotation.

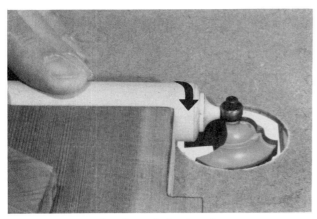

Illus. 24-19. Forming the end of a dowel.

Illus. 24-20. Various cuts made with this one dishing cutter used in an angled router.

Illus. 24-21. Approaching the bit. Note the V-grooved push block, the bevelled stop, and the end fence lateral stop.

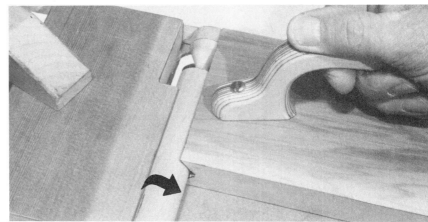

Illus. 24-22. Making the cut. The dowel is turned by hand in the direction of the arrow.

The Up-Plunge Router Table

The upcut plunge router table (Illus. 24-23) looks rather crude, but it works great. It makes almost every conventional router table operation safer. In the same sense that a hand-held plunge router extracts the bit at the completion cut, when this table is used the bit automatically disappears below the table with the release of the foot pedal. This is excellent for routing stopped grooves and mortises of all kinds which normally can only be cut by dangerously lowering the workpiece over a rotating bit set at a fixed height.

Illus. 24-23 (right). The author's prototype of the up-plunge router table.

The router is bolted directly to a flat plywood top ¾″ thick (Illus. 24-24). No router mounting plate was used for this particular model, but one could be—provided it is rigid and it is bolted to the top. Most routers have a 2″ or greater stroke capacity, which still provides a 1¼″ depth of cut above the table, sufficient for cutting many mortises. The big Elu router has almost a 2½″ stroke, which is more than sufficient for most table work.

The up-feeding mechanism is such that pressure is applied against the top of the motor housing, which works very well for most plunge routers. The design could easily be modified to connect the lift arm directly to the handles of the routers, but each brand of router is slightly different.

All of the vertical framing members are 1½″ × 2½″ pieces rip-sawn from 2 × 6 cedar planks. Illus. 24-25–24-28 show and describe the structural components of this new and safer routing system developed for all woodworkers.

Illus. 24-24. A pair of polypropylene (or plywood) pivot pads on a riser bar presses the router upwards as the foot pedal is depressed.

Illus. 24-25. The depth of cut is controlled by the router's own adjustment mechanism. On this Elu router, the height depth stop can also be adjusted to limit the vertical stroke capacity.

Illus. 24-26. This view without the top shows the very basic structural details of the router table.

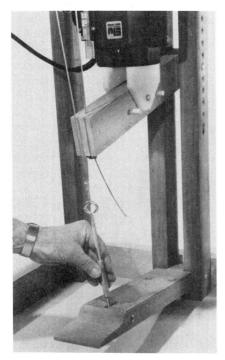

Illus. 24-27 (left). A turnbuckle adjustment keeps the foot pedal a comfortable working height from the floor. Illus. 24-28 (above). The up-plunge router table is ideal for all template-guided jobs, but especially those that require routing inside openings. The bit can plunge up into the work with the work held flat on the table. Here an outside trim cut is being made with a pin attachment that is following a vacuum template.

Joint-Making Techniques with a Hand-Held Router or a Router Table

25
Biscuit Joinery

Biscuit (also called plate) joinery has become increasingly popular in recent years. Today, special power tools called biscuit joiners are available. Essentially, all they do is cut small, matching curved slots into components of wood joints so that the biscuits—thin, football-shaped splines—can be inserted into the slot. The biscuits swell tight as they absorb moisture from the glue. Because biscuit joinery is easy, fast, and the layout is very simple, it has become the alternative to tenon, tongue, spline, and dowel joints.

The router, equipped with a ball-bearing-guided two- or three-wing slot cutter $1\frac{7}{8}''$ in diameter that cuts a $\frac{5}{32}''$ slot, will produce perfect slots for all biscuit sizes. You can cut biscuit slots in three basic ways with your router: (1) hand-held, freehand, where you hold the base flat on the surface and slide the router to make a horizontal plunge cut; (2) hand-held, using the Sears Bis-Kit system; and (3) in a router table.

Sears Bis-Kit System

The Sears Bis-Kit system (Illus. 25-1 and 25-2) is designed for most Sears routers. The standard sub-base is replaced with one that has a short inward-feeding travel-carriage mechanism that allows you to plunge the cutter into an edge of the workpiece after you align the accessory to location marks on the wood. You have to slide the router along the edge a short distance specified by indexing marks on the attachment for the lengths of various-size biscuits.

Illus. 25-1. The Bis-Kit system attaches to almost all Sears routers with 6″ bases with just three screws.

Illus. 25-2. The Sears Bis-Kit system in use. Locating marks on the wood are aligned with the accessory. A short inward and side feed stroke cuts the slot for the biscuit.

Using a Router Table

Making biscuit joints with a router table is very easy. You can cut all of the same kinds of edge slots an expensive biscuit joiner will cut, but you cannot make cuts into the surface of a board with the router table to connect a shelf to vertical panels.

The joints shown in Illus. 25-3–25-8 were all made on the router table. Lay out the location for each biscuit.

Draw a centerline across the surfaces of both joint components near their edges. Install a 5/32″ slot-cutting bit and set it to its appropriate height. Set the fence to expose the bit to the desired horizontal cutting depth.

Draw two lines on the fence, one on each side of the center of the bit. Their location is determined by the length of feed required for the size biscuits you are using (Illus. 25-9 and 25-10). Start the cut at the right or near

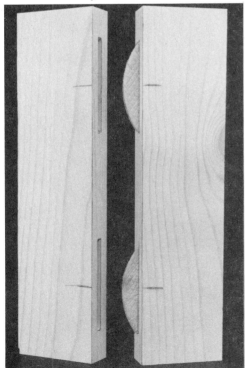

Illus. 25-3. Typical edge-to-edge biscuit joint.

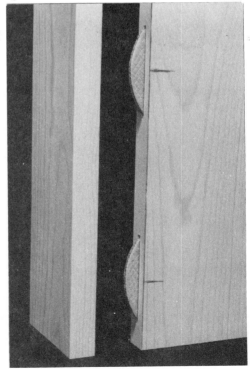

Illus. 25-4. An edge-to-face butt joint.

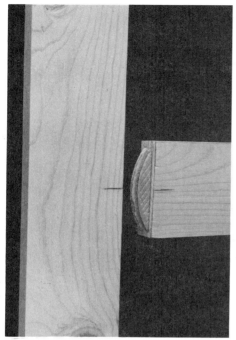

Illus. 25-5. A rail-to-stile joint.

Illus. 25-7. Flat mitre joints.

Illus. 25-6. An end-to-face butt joint.

Illus. 25-8. An edge mitre joint.

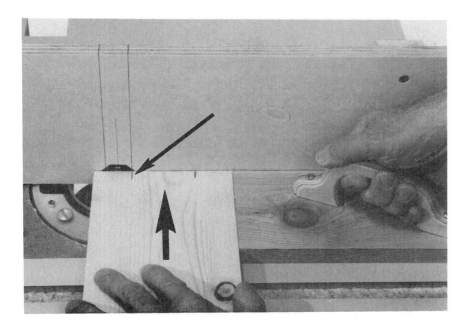

Illus. 25-9. The setup for cutting the first of two slots into the end of a board. To begin the cut, feed the work into the bit so the centerline of the biscuit lines up with the right vertical line drawn on the fence. Note the push block used to guide and advance the work.

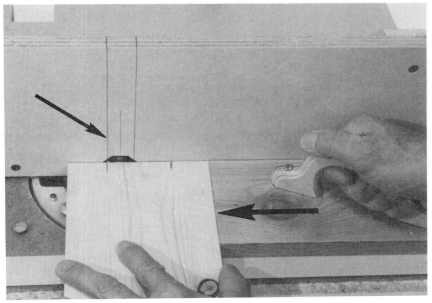

Illus. 25-10. Once the workpiece contacts the fence, advance it to the left reference line with the push block.

the side mark and advance the workpiece until the centerline of the workpiece is aligned with the second (left) mark on the fence.

All cuts must be made by pivoting or feeding the workpiece horizontally into a rotating bit. You can use a corner or edge of the workpiece as a fulcrum point or use a stop clamped to the fence or table. All prudent safety provisions should be employed. Always keep your hands at a safe distance and be sure to use appropriate fixtures when cutting slots in smaller workpieces.

Illus. 25-11–25-15 show how to feed the workpiece into the rotating bit when making various joints. Other, more appropriate methods can be employed to specifically handle certain cuts, workpiece sizes, and production needs.

Illus. 25-11. Pivoting the workpiece to cut slots for an edge-to-edge biscuit joint.

Illus. 25-12. Pivoting a mitred piece into the slotting cutter to make flat-mitre biscuit joints.

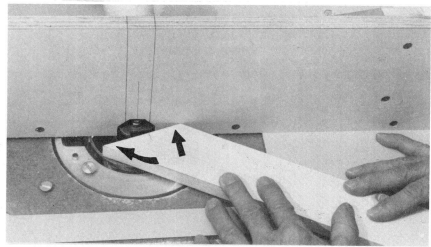

Illus. 25-13. Pivoting a workpiece into the slotting cutter to make slot cuts on the end of the workpiece's face. Reference lines drawn on the back surface of the work and on the top of the fence are aligned, for starting and stopping the cut. Note the stop block clamped to the table.

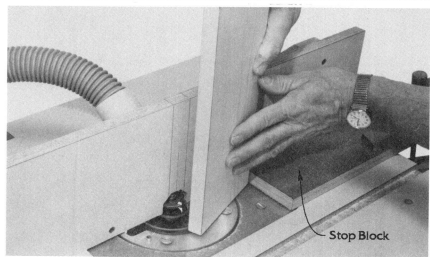

Stop Block

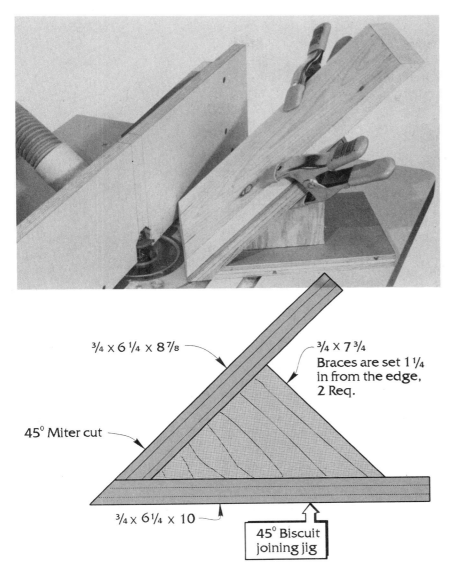

Illus. 25-14. A simple jig supports the work at 45 degrees for slot cuts on the edges of mitres.

$\frac{3}{4} \times 6\frac{1}{4} \times 8\frac{7}{8}$

$\frac{3}{4} \times 7\frac{3}{4}$
Braces are set $1\frac{1}{4}$ in from the edge, 2 Req.

45° Miter cut

Illus. 25-15. A side view of the jig.

$\frac{3}{4} \times 6\frac{1}{4} \times 10$

45° Biscuit joining jig

26
Dovetails

Sharply executed and perfectly fitting dovetail joints are unquestionably the hallmark of high-level craftsmanship (Illus. 26-1 and 26-2). Dovetail joints attract the eye. Woodworkers scrutinize the quality of the joints and pass judgment on the maker's overall woodworking capabilities. This chapter will help you become a dovetailing expert.

Illus. 26-2. A half-blind dovetail joint.

Illus. 26-1. A through dovetail joint produced with a hand-held router and a commercial template jig.

Bits for Dovetailing

As discussed earlier, bits come in a variety of sizes. When ordering bits, specify the shank size, the angle desired, the largest cutting diameter, the cutting height, and whether you want an hss- or carbide-tipped bit (Illus. 26-3). Bits are available in cutting diameters from 5/16″ to 1¼″ and cutting heights of ¼″ to ⅞″. Illus. 26-4 shows two dovetail cuts with different slope angles. Notice that the more inclined the slope, the more brittle the corners become because of short grain. The manufacturers of commercial fixtures usually recommend the bits needed. Big bits do require routers with ½″ collets.

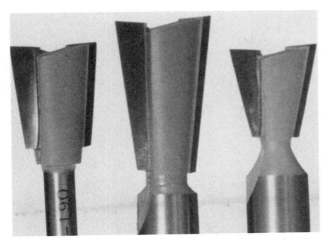

Illus. 26-3. Typical dovetail bits of various slope angles. From left to right are bits with 7-, 8-, and 14-degree cutter angles.

Illus. 26-4. Give some thought to the slope angles of bits and the resulting dovetail design. Higher angles are more subject to splintering, as shown.

Dovetailed Dadoes and Grooves

Using the router is the only practical way to make long, straight-line dovetail cuts in any direction. Such cuts, with matching dovetailed tongues, are used for panel joints in fine furniture and cabinetry. It is relatively easy to figure out how to cut these joints. Any of the router-guiding methods discussed in previous chapters can be used: employing straightedges, T-squares, frame guides, or fence guides on a router table. *Note:* To minimize the possibility of the bit getting trapped in a chip-impacted dovetail groove, preplow the cut with a straight bit to remove most of the waste.

The open-ended mortises, or slots, in leg-and-rail construction used on tables cut with a dovetail bit and fitted with matched dovetails on the tenons make very strong joints. Joints of this type are used for some knock-

down furniture, and household accessories. All such joints are easiest to make on the router table. However, Illus. 26-5 and 26-6 show how male and female dovetail cuts can be made using the portable router with an edge guide.

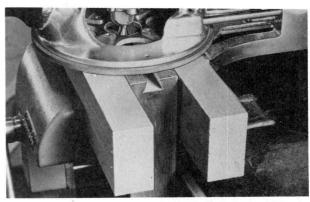

Illus. 26-5. Making the female dovetail cut. Note that blocks are clamped to each side, to give it support.

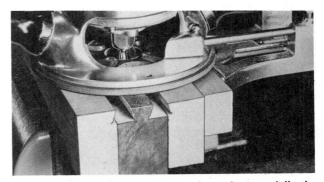

Illus. 26-6. The male dovetail tenon is cut in essentially the same way.

Sliding Dovetails

Sliding dovetails have a number of applications of interest to the creative woodworker. Obviously, the fit of the matching members is less exact, with sufficient clearance intentionally provided to ensure smooth movement of the mating parts. A few items on which sliding dovetails can be used include box covers (Illus. 26-7 and 26-8), drawer slides and glides, adjustable shelves, adjustable track lighting, adjustable book-and-record holders, and various adjustable tools and work-guiding fixtures used for woodworking.

Illus. 26-7. Boxes totally router-machined. The cuts for the sliding lids were made on the router table.

Illus. 26-8. These dovetail boxes by the author have sliding lids.

Dovetailed-Keyed Mitre Joint

This dovetailing technique (Illus. 26-9), widely used with plywood panel construction, simultaneously provides decoration, strength, and creativity to the conventional edge mitre joint. First, cut and glue the mitre joint in the usual manner and allow the glue to dry. Then rout the dovetail key cuts across the assembled joint. There are several ways to do this. You can devise a slotted guide with a V-block fixture that cradles the joint to support and guide the portable router. Another way is to make the key cuts on the router table. Use the fence to guide the work while it is held in a supporting cradle fixture.

Commercial Dovetail Jigs and Fixtures

Commercial dovetail jigs and fixtures are available to make the two most popular dovetail joints: half-blind and through. Most jigs do come with understandable instruction manuals, and some even come with demonstra-

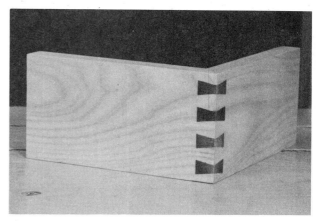

Illus. 26-9. The dovetail-keyed mitre joint.

tion videos, but sometimes the manufacturers assume that you know all the dovetailing jargon. The through dovetail can be seen on the two surfaces of the joint; the half-blind dovetail can be seen only on one side. Half-blind dovetails are primarily used for flushed and lipped drawers.

Tails and pins are the two different parts of a dovetail joint. Illus. 26-10 shows both parts. Tails taper with the face grain. Pins taper across the end grain.

Dovetail Spline Jig

The dovetail spline jig (Illus. 26-11 and 26-12) is a plastic-template fixture measuring $2'' \times 3\frac{1}{2}'' \times 12''$ that clamps to a preglued mitre joint. The jig guides dovetail

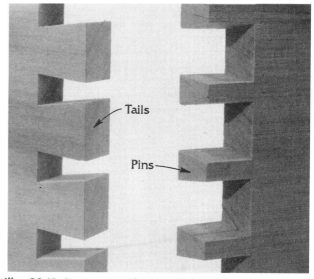

Illus. 26-10. Components of a through dovetail. Tails are shown on the piece on the left. Pins are shown on the piece on the right.

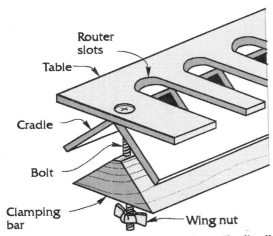

Illus. 26-11. The parts of the Kehoe dovetail spline jig.

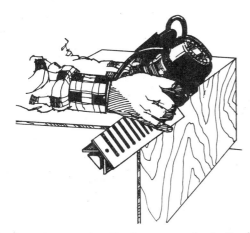

Illus. 26-12. The Kehoe jig clamps to a glued mitred corner. To use it on work longer than the normal capacity of the jig, drill a hole through the joint to clamp the jig down. The hole is later covered with a spline insert.

Illus. 26-13. The keys of splines are of solid wood and contrasting colors. They are cut slightly longer, glued in, and then sanded flush.

Illus. 26-14. Half-blind dovetails are cut with a commercial finger template fixture to make flush (left) and lipped (right) drawers.

cuts across the assembled corner. A template guide mounted in the base controls each cut. Individual keys (or splines) of contrasting wood are fitted and glued into the dovetailed channels (Illus. 26-13). Then they are cut and sanded flush. The Kehoe jig is available from L. C. Kehoe Mfg. Corp., Whitefish, Montana.

Fixtures for Half-Blind Dovetails

Fixtures for half-blind dovetails (Illus. 26-14 and 26-15) have been around for many decades. Those currently on the market are much improved over their earlier counterparts. When they are set up properly, it's possible to cut dovetails all day, making a complete joint every couple of

minutes. Setting up the fixtures for dovetailing does require time, patience, and some trial-and-error work to get the bit depth, shims, stops, template location, and similar adjustments all properly set.

Some companies that provide half-blind template fixtures include Black & Decker, Bosch, Sears, and Porter-Cable. Tests indicate that they all cut good dovetails, but some require substantially more setup time than others. Prices vary, as do construction features and size capacities. Their board-width capacities vary upward from 12″.

Most router-dovetailing fixtures for making half-

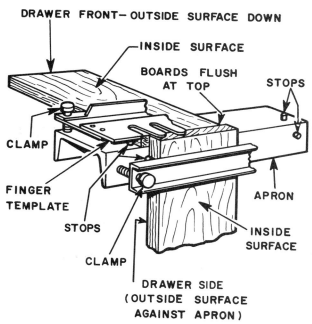

Illus. 26-15. The essential parts and setup for routing half-blind dovetails. Note that the pieces are routed with their inside surfaces facing out.

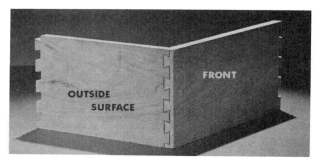

Illus. 26-16. Two pieces assembled for a drawer with a flush front.

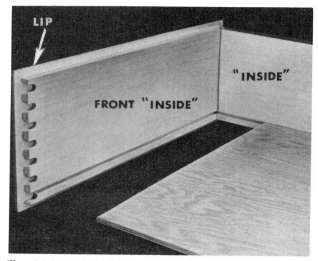

Illus. 26-17. Details showing a lipped drawer front. Note the dovetail pins cut into the front.

blind dovetails come with a standard, plastic finger template and are designed to machine ½″ dovetails, the largest size that can be cut. These fixtures are used to cut dovetails in stock from ⁷⁄₁₆″ to 1″ thick. Also be sure to use a router-base-mounted template guide (bushing) that has a ⁷⁄₁₆″ outside diameter and a ½″ dovetail bit with a ¼″ shank. This setup is used to cut dovetails in the most popular sizes for drawers for furniture and cabinets (Illus. 26-16 and 26-17). Refer to the owner's manual for specific instructions. For example, various guide pins have to be set when you are changing from flush- to lip-drawer dovetailing. If the pieces fit too tightly, the bit needs to be raised. If they fit too loosely, the bit should be lowered. The finger template is also adjustable in or out, to correct cuts that are too shallow or too deep. All of this requires some time spent on trial-and-error adjustments and test cuts.

Incidentally, it's also possible to purchase an optional, second interchangeable finger template for routing smaller, ¼″ dovetails in thinner stock (⁵⁄₁₆″ to ⁵⁄₈″). This is used with a router-base-mounted template guide (bushing) that has a ⁵⁄₁₆″ outside diameter and is used with a ¼″ dovetail bit.

The really appealing feature of these fixtures is that once set up and adjusted properly, both the pins and tails

of the joint are machined at the same time. Both members of the joint are clamped together on the fixture. The router is carefully moved in a left-to-right direction in and out, between and along the fingers of the template (Illus. 26-15), cutting the pins and tails simultaneously in one operation.

One special word of caution: It is so easy to inadvertently lift the router when completing a cut, because this is a normal router movement in many other router operations. Don't do this when dovetail-routing with all fixture templates, because you will strike the finger template with the rotating bit and completely ruin it. The router must be slid away in a perfectly level and horizontal direction. It's best to shut down the router, wait for the bit to stop revolving, and then slide the router off the template without lifting it.

In spite of the setup time spent when using these

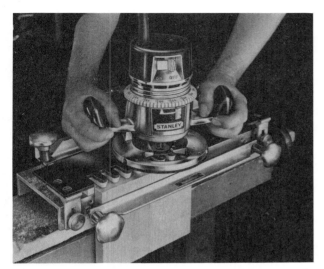

Illus. 26-18. Both members of the joint are clamped into the fixture (one vertically and one horizontally). Carefully follow the template fingers. Feeding left to right and then back again ensures that all cuts are clean and complete.

fixtures, the end results are well worth the trouble. The finished dovetails will fit perfectly, and they are strong, durable, and professional-looking joints.

Leigh Multi-Functional Jigs

The Leigh multi-functional jigs (Illus. 26-19 and 26-20) come in work-width capacities of either 12″ or 24″. The Leigh jig is the invention of Ken Grisely, President of Leigh Industries in Port Coquitlam, British Columbia,

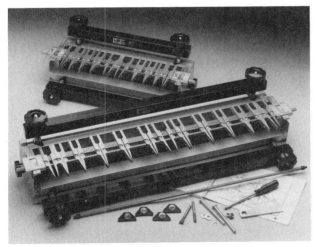

Illus. 26-19. Leigh dovetail jigs are available in width-cutting capacities of 12″ and 24″.

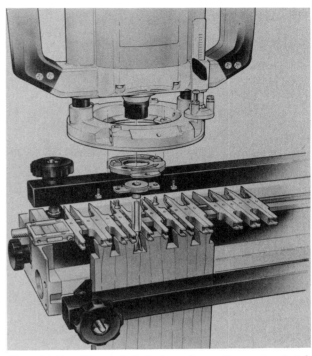

Illus. 26-20. This exploded view shows the router, Leigh adapter, template guide, bit, and workpiece in the jig for through-routing in the tail-cutting mode. Not shown here are new cam-action clamps available for the jig.

Canada. The Leigh jig can be used to cut through and half-blind dovetails of any size or spacing. Additionally, it can be set up to make round-cornered through or blind mortises and tenons. Also, with the Leigh jig you can use a hand-held router to cut a variety of unusual joints in pieces that are too big to handle on the router table.

The one advantageous feature of this jig is its capability to produce variable-spaced dovetails (and mortise-and-tenon cuts) where you want them. In other words, cuts do not need to be uniformly spaced from each other, as is the case with almost every other jig. The variable-spacing design incorporated in Leigh's adjustable template fingers makes layout really easy. Simply install the wood in the jig and then space the fingers for the cuts where you want them (Illus. 26-21).

The reversible finger assembly on the Leigh jig creates the routing patterns for both the pin and tail cuts of the joint. One end of each finger is pointed with an angular face (for pins), and the other end is square or with a straight guiding face (for tails).

When using this jig, first make all the cuts necessary

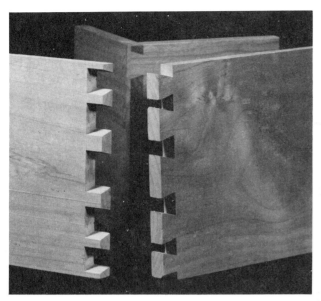

Illus. 26-21. A sample of variable spaced dovetails cut with the Leigh jig. The tails are on the right, and the matching pins are on the left.

in one mode (tails or pins), and then flip the finger template assembly to access the other mode.

Straight-cutting bits are used for the pins, and dovetail bits for the tails. The bits are used with a router-base-mounted template guide to follow along and in and out of the fingers. The workpieces are cut while held vertically in the jig. Both pieces are not cut together simul-

taneously, as they are in other jigs designed just for half-blind dovetailing. One pass must be made with the router (carrying a dovetail bit) to cut the tails (Illus. 26-20). Another router setup and separate pass must be made using a straight-cutting bit, to cut the pins (Illus. 26-22). Having two routers makes the job much more convenient—especially for small production runs. Set one router up for tail-cutting, and another for pin-cutting.

The Leigh jig is a very sophisticated piece of equipment and offers a wide range of joinery capabilities for the creative router craftsman. In addition to its mortising and tenoning capabilities (Illus. 26-23), with some im-

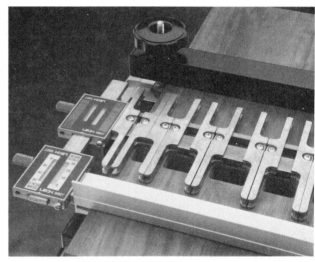

Illus. 26-23. The Leigh jig in its "mortise-cutting" mode with the multiple mortise-and-tenon attachment. Use it to cut one or a group of through mortise cuts, as shown here.

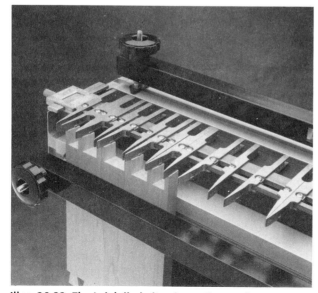

Illus. 26-22. The Leigh jig in its pin-routing mode.

Illus. 26-24. The Leigh jig was used to cut the dovetails in the angular drawers of this unusual desk by Peter Van Herk of Gloucester, New South Wales, Australia.

Illus. 26-25. This decorative, two-stage double half-blind dovetail was made by Ken Grisley with his Leigh jig.

provision it can be used to make functional or decorative dovetails on angular cut corners (Illus. 26-24). It can even produce some versions of the new, popular double half-blind dovetail joints (Illus. 26-25). The jig comes with a complete and detailed instruction manual written by woodworking expert Bill Stankus.

Porter-Cable Omnijig

The Porter-Cable Omnijig (Illus. 26-26) permits production-routing of half-blind, through, sliding and other dovetails plus box joints. The unit has a 24″ workpiece-width capacity. It comes with a template for making half-blind dovetails, a template guide, bit, and an instructional video.

Illus. 26-26. The Porter-Cable Omnijig has a 24″ stock-width capacity. It takes a variety of optional fixed-finger and adjustable templates.

A variety of special, interchangeable templates are available (Illus. 26-27), including one for making adjustable through dovetails. One of the strongest features of

Illus. 26-27. Here are some of the interchangeable templates available for the Omnijig. Note that some are in pairs and most are nonadjustable and machined from ¼″ aluminum. The adjustable through dovetail template shown on the far right has fingers (guide forks) of solid ½″ aluminum extrusions that slide on steel rods and can be locked with thumbscrews.

this jig is its heavy cast construction; it weighs 55 pounds. It also has a one-hand cam-action clamp of 1¼″ steel that is strong enough to remove any cupping from the workpiece being routed. You can purchase special templates as desired to make a variety of standard joints (Illus. 26-28).

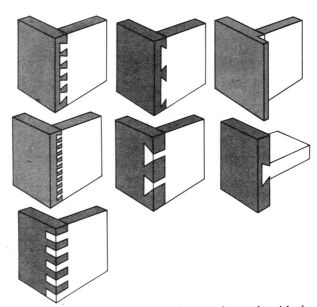

Illus. 26-28. The dovetail joints that can be made with the Omnijig. Top left to right: ½″ half-blind, 2-inch-spaced, and rabbeted dovetails for lip drawers. Center, left to right: ¼″ half-blind, through (adjustable-spaced), and sliding dovetails. Below: ½″ box joint.

Keller Dovetail Templates

The Keller dovetail templates (Illus. 26-29) are used for through dovetailing in cabinet and furniture construction. This dovetailing system is amazingly simple, straightforward, and very accurate. They are designed, manufactured, and distributed by David A. Keller, Petaluma, California. The model 3600 dovetail templates with shank ball-bearing-guided bits (Illus. 26-30) will cut standard, true through dovetails in stock up to 1¼″ thick. There are two smaller models. The 16″ model will handle stock up to ¾″ thick, and the 24″ model can be used to cut stock 1″ thick.

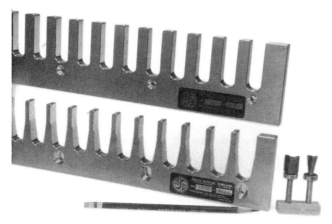

Illus. 26-29. The Keller® dovetail templates come in sets: one to cut tails (above), and one to cut pins (below). They are made of precision-cut ½″ aluminum and are available in lengths of 16″, 24″, and 36″.

Illus. 26-30. Shank-mounted, ball-bearing-guided bits match perfectly with the template slots.

The initial setup is relatively easy, and you can control the tightness of the templates by positioning them appropriately on mounting blocks made by the operator. Keller's dovetail templates are easy to set up and use. The pin spacings are obviously all nonadjustable. The pins are spaced 1⅛″, 1¾″, and 3″ from center to center, respectively, on the 16″, 24″, and 36″ sets.

Optional variable spacing can be achieved by shifting the templates after each cut, but keeping the alignment is difficult. The templates are big enough to handle almost any joint width. The biggest model will handle stock up to 36″ wide at one setting, but the capacity of each model can be increased infinitely by shifting the templates along the work. Illus. 26-31–26-33 show close-ups of the process involved in making the tail and pin cuts.

Refer to Chapter 27 for information on cutting boxed joints using the Keller dovetail template.

Illus. 26-31. The operator's view of routing the tails.

Illus. 26-32. Close-up of the tail-cutting action. Note the soft feathering on the out-feed side, which is easily removed with one pass of fine-grit abrasive.

Illus. 26-33. Routing the pins with the Keller template.

Cutting Dovetails Without Commercial Jigs or Fixtures

This section explores an alternative procedure to using commercial jigs or fixtures for through dovetailing. It is just one of numerous ways dovetails can be cut, and can be modified to your own preferences. This method combines basic hand-held and table routing techniques, so that a minimum of hand-tool work is actually necessary.

The procedure given here can be applied to making dovetails in stock of any thickness with any pin and tail spacing. It generally enables anyone to design and produce perfect pin and tail cuts on workpieces, for professional-looking joints of any width. The resulting joints will look so good they will appear to be made with the expensive jigs described earlier.

Although the total procedure may appear at first to be long, involved, and tedious, it really isn't. Once you have made the necessary fixtures and practised on two or three joints, you will be able to quickly make the dovetail joints.

The illustrations in this section show large dovetails being cut in ¾″ stock using a large bit with a ⅞″ cutting height capacity. This bit can be used to cut various through dovetails in stock from ½″ to ⅞″ thick. The process can be applied exactly the same way with smaller bits intended for dovetailing thinner stock. The cutting height of the dovetail bit determines the maximum thickness of stock that can be cut.

First, design and lay out the tails of the joint directly onto one piece of work. Simply make a paper or template pattern of a pin and use it to draw the tails wherever

desired (Illus. 26-34). A clear-plastic template is easier to use. The joint should be laid out so the spacing between the tails is at least ⅛″ wider than the cutting diameter of the dovetail bit. The tails will be cut on the router table. With the dovetail bit installed, set the cutting depth exactly equal to the thickness of the pin workpiece (Illus. 26-35). All the tail cuts will be made first, before cutting any pins.

Illus. 26-34. A wood template conforming to the size and profile of the dovetail bit is being used to design and lay out the pin openings on the tail piece.

Illus. 26-35. With the bit in the router table, set its height to equal the stock thickness plus slightly more.

Set the fence (Illus. 26-36) to locate the first cut so it is made approximately the desired distance from the edge

Illus. 26-36. Set the fence approximately to the distance desired for
making the first cut.

of the workpiece as initially designed and laid out. The
fence does not have to be precisely located. As many
identical pieces as can be conveniently clamped together
should be cut at the same time. This operation requires a
support fixture similar to the one shown in Illus. 26-37
and 26-38. An auxiliary backup board is inserted between the workpieces and the support fixture to eliminate any splintering or tear-out as the bit exits from the
work.

Illus. 26-39 shows an alternate mitre-gauge dovetailing setup using an auxiliary fence with a small work-
positioning stop. This arrangement will produce evenly
spaced dovetail cuts. The stop is made to the same size as

the narrow end of the dovetail cut. After the first dovetail
is cut, the next and successive cuts are made by setting
each preceding cut over the stop.

Complete the routing of all of the tail pieces on the
router table. All the tails should be cut to identical depths.
Identical tail widths are not critical. Thus, tails can be
large (wide) or small (narrow), as desired.

Next, cut the sides of the pins with a handsaw or scroll
saw. The location for each cut will have to be marked onto
the work (Illus. 26-40). Cutting the pins is tricky, but
certainly not difficult. There will be less confusion during cutting if "hash marks" are used to identify the waste
areas where stock must be removed (Illus. 26-41). A

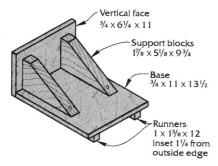

Illus. 26-37. Making a test cut. Note the
construction and function of the
support-slide jig.

Illus. 26-38. Making the cuts. The sliding fixture is guided against the fence. The board backer behind the two work-pieces eliminates splintering as the bit exits from the cut pieces.

Illus. 26-39. A mitre-gauge setup for making equally spaced dovetail cuts. The auxiliary facing board with a work-positioning stop is fastened to the mitre gauge.

Illus. 26-40. Use a very sharp pencil and lay out all the pin cuts using a cut tail piece as a pattern.

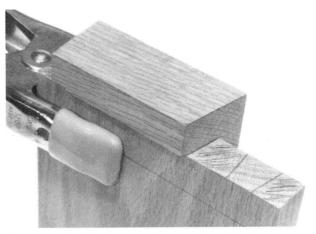

Illus. 26-41. The block clamped to the work guides the dovetail saw. An accurate block is essential. It must have the same angle as the dovetail bit used. Clamp it so it just covers the layout line. Note the "hash marks," which clarify which areas need to be cut away.

simple but carefully made guide block will help with hand-sawing the sides of the pin (Illus. 26-42). It's best to make the guide blocks with a table saw, for absolute accuracy. The angle on the block must be exactly equal to the angle cut by the dovetail bit.

To ensure that the joints will not fit too tightly or loosely when assembled, clamp the block precisely to the work so it just covers the layout line. The vertical cuts can now be made with a dovetail or other fine-toothed saw (Illus. 26-42 and 26-43).

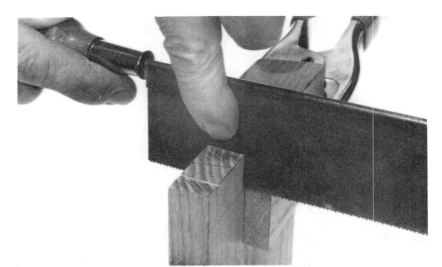

Illus. 26-42. The saw is held against the guide block to make true, vertical cuts. The block can also be made to function as a depth stop for the saw, eliminating layout lines.

Illus. 26-43. A preferable alternative for cutting the sides of the pins is to use a scroll saw with a tilted table to follow the layout lines.

Illus. 26-44. Adjusting the bit depth for routing the pins.

The next step is to remove most of the waste between the pins and make a true and finished bottoming cut. Both of these jobs can be handled simultaneously with the portable router and a straight-cutting bit. Install a straight-cutting bit and adjust its depth of cut so it equals the thickness of a previously made tail-cut member plus slightly more (Illus. 26-44).

Two or more pieces should be clamped together for the pin-routing operation (Illus. 26-45). An extra square and flat-edged board should be clamped flush to the work surfaces to help support the router base. Remove the waste material between the saw kerfs using freehand-routing techniques. Stay safely away from the pin stock. If it feels more comfortable, cut only to within ⅛″ or ¹⁄₁₆″ of the pin's saw kerfs (Illus. 26-46 and 26-47).

Use a sharp knife to remove any remaining waste (Illus. 26-48). With some practice using the router, it is possible to freehand-rout into the kerf and still exercise sufficient router control so the bit doesn't touch the sawed surfaces of the pins.

Make a test fit of the mating joint pieces. They should not require much force. Once they have been assembled, glued, and clamped, you should feel a little protrusion of the tails and pins. These can be sanded flush or simply taken off with a flush-trimming bit. Remember, cutting pins and tails slightly longer is far better than having them too short with their ends below the surface.

Boxes, cases, and similar projects can all be made employing the above router-assisted, hand-fitted dove-tailing procedure. This method of making dovetails is faster than one that involves all hand tools, and is much more accurate and reliable. Furthermore, it's gratifying to the ego (and pocketbook) to know that dovetailing can be done without the use of expensive, commercially made jigs.

Illus. 26-45. Two pin pieces with pre-sawn vertical kerfs held together in a vise. Clamp an extra board flush, to provide additional support for the router.

Illus. 26-46. The operator's view of freehand-routing the waste between the pins.

Illus. 26-47. The freehand waste removal completed. Note that cuts must be made into the support piece and that the cuts are made only fairly close to the pins.

Illus. 26-48. Some quick slicing strokes with a sharp knife will clean up the cuts.

Illus. 26-49. The assembled joint. The pin and tail ends should extend slightly above the surrounding surface.

27
Box Joints

Box joints (Illus. 27-1) are also called "finger joints" in the United States and "comb joints" by the British. Like dovetails, they can be router-machined in a variety of ways. Commercial devices include the Porter-Cable Om-nijig, the Keller 16″ dovetail template, and the Hegner finger-joint machine. You can also make and utilize your own fixtures on the router table with which to cut box joints.

Box joints are not as widely used overall as are through dovetails. Box joints are not as strong or as interesting to look at, and they require almost as much effort to set up and cut as dovetails.

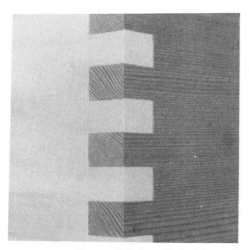

Illus. 27-1. A box joint cut with the Keller 16″ dovetail template.

Keller 16″ Dovetail Template

The Keller 16″ dovetail template has a special shank-mounted ball-bearing-guided bit that matches the finger slots. The setup is extremely quick. It's easy and permits the cutting of both pieces of the joint at one time. Simply center one member of the joint under the template fingers so that the spaces or fingers are equal at the edges.

Offset the edge of the second board ⁹⁄₁₆″ from the edge of the first piece and clamp the jig to both boards in this position (Illus. 27-2). The bit required is a ⁹⁄₁₆″ diameter straight-cutting bit with a ⅝″ outside-diameter bearing. Cut both members of the joint simultaneously (Illus. 27-3).

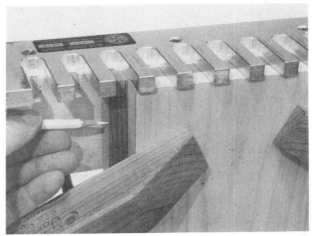

Illus. 27-2. Offset the workpieces ⁹⁄₁₆″, as shown, regardless of what thickness they are.

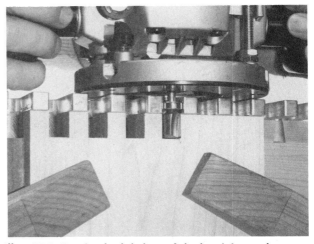

Illus. 27-3. Routing both halves of the box joint at the same time.

Using a Router Table

Making box joints with a router table (Illus. 27-4) is done as described on pages 313–318, which show how to cut uniformly spaced dovetails. A properly located indexing guide block locates the workpiece relative to the bit (Illus. 27-5 and 27-6). All the resulting cuts are perfect and equally spaced.

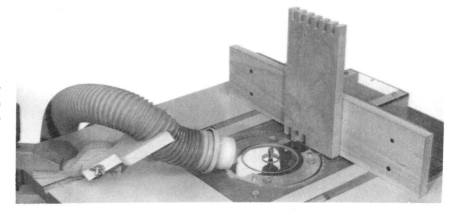

Illus. 27-4. The router table set up for cutting box joints and dovetails. Two strips ride in slots cut in the table to guide this work-holding fixture.

Illus. 27-5. An indexing block positions the work for the next cut. The bit diameter and pin-width setting must be exactly equal.

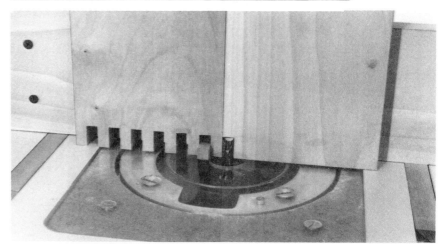

Illus. 27-6. Start the first cut on the second piece of the joint while this piece is butted next to the first piece, as shown.

Hegner Comb-Jointing Machine

The Hegner comb-jointing machine (Illus. 27-7) is a German-made product dedicated to cutting box joints without templates. This moulded plastic device clamps to the edge of a workbench. It has a platen or carriage that feeds vertically supported stock back and forth across a fixed bit (Illus. 27-8).

Stock of any width and length and up to approximately ⅞″ in thickness can be milled on this machine. The workpiece(s) is (are) just hand-held against a flat upright face. The size of the cut(s) is determined by the diameter of the bit used.

A pointer on the front (operator's side) and a side-adjustment knob control a work-indexing finger that's precisely set according to the size of the bit (Illus. 27-9 and 27-10). The power source is an AEG, 27,000 die-grinder motor unit. I've been told that a Bosch laminate

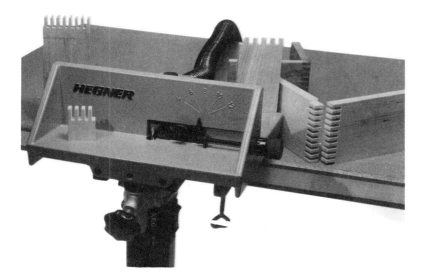

Illus. 27-7. Hegner's comb- or finger-jointing machine cuts box joints without templates. Note the pointer and microadjustment knob for matching bit diameters designated in millimetres.

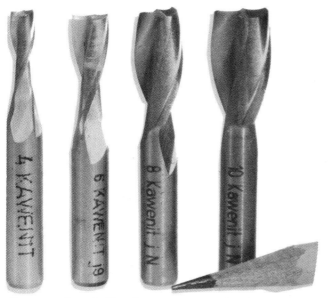

Illus. 27-8. The spiral bits for the Hegner machine have cutting diameters of 4, 6, 8, and 10 millimetres. An optional 12-millimetre bit is available.

Illus. 27-9. A close-up showing the pointer (set to 4 millimetres) that sets the stop to match the bit diameter.

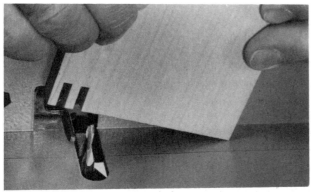

Illus. 27-10. A cut about to be made. This view of the workpiece side of the vertical support shows the adjustable work-positioning stop in relation to the bit. The guard has been removed, for photo clarity. Always use the guard of your machine.

trimmer will also fit the 43-millimitre mounting collar.

A connector for a vacuum chip-extraction hose is provided, along with a transparent guard that's adjustable to stock thickness, so the bit is always covered. The guard on the Hegner machine illustrated in this section has been removed, so that there would be a clear view of the machine. You should always use the guard.

Illus. 27-11 shows a completed joint produced with this product. The Hegner finger joint jig is available in the United States from Advanced Machinery Imports of New Castle, Delaware.

Illus. 27-11. A completed finger-joint corner. Note the small size of the fingers and the perfect fit.

Shop Aids and Accessories, Specialized Routing Devices, and Routing Machines

28
Precision Fences

The spectacular decorative joints (Illus. 28-1 and 28-2) and accurate router-table cuts produced with the aid of the relatively new precision-routing fences (Illus. 28-3 and 28-4) have captured the interest and curiosity of most woodworkers. Essentially, these devices are fences with mechanisms capable of adjustment to within .001″; the mechanisms will then reposition the fence to the exact-same location when needed. The adjustments are made without measuring.

Included with the precision fence is a right-angle feeding device (Illus. 28-3) and various adjustable stops that combine to control very accurate spacing from the fence to the bit and the length of cuts. This provides precise routing of dadoes, rabbets, tongues, slots, slits, grooves and dovetails in straight-line directions (Illus. 28-5). In short, these are more than just fences; they are a machining system.

Two companies manufacture and market products that essentially do the very same things. This chapter will present a very brief overview of the features on these systems and how these systems are used. Specific directions pertaining to setup, adjustment, and use would require lengthy discussions, and this information is detailed very well in the manufacturer's manuals, booklets, and demonstration videos.

The first precision fence to appear was the Incra Jig, introduced in 1987 by the Taylor Design Group of

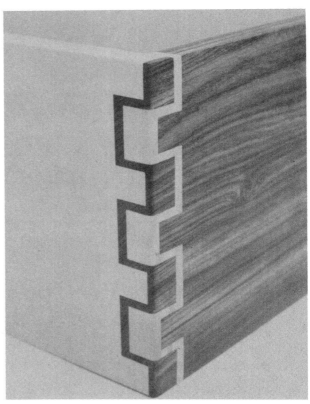

Illus. 28-1. A close-up of a typical Incra double-double box joint originated by Christopher Taylor.

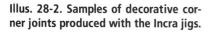

Illus. 28-2. Samples of decorative corner joints produced with the Incra jigs.

Illus. 28-3. The Incra Pro precision-positioning jig mounted to a router table. Note the fence, fence stop, and the right-angle sliding fixture that hangs over the fence. Also note the configuration of the cut in the table for the router-mounting plate.

Illus. 28-4. JoinTECH's incremental positioning fence mounted to a router table.

Dallas, Texas. It was promoted as the Universal Precision Positioning Jig (Illus. 28-6 and 28-7). Within the past few years JoinTECH Corp. of San Antonio, Texas, introduced its version, called the Incremental Positioning Machine. And just recently Incra countered with Incra Pro (Illus. 28-3), a larger, improved version of the original Incra jig.

There are structural and operational differences between the Incra and JoinTECH systems. Costs should be compared, but if you are buying be sure you also compare fence adjustment range, fence length, accessories such as stops, right-angle fixtures, hold-downs, templates, and the usefulness of the instructional material.

The fences that come with these systems are similar. They are made of flat and straight extruded aluminum (Illus. 28-6). The square clearance openings for bits cut into the fences are only 1″ by 1″, which indicates that these systems are not intended for big, heavy cuts. However,

Illus. 28-5. The capability of setting up and machining perfectly matching sliding dovetails such as these is the same general concept involved to produce the spectacular double, double corner dovetails and box joints.

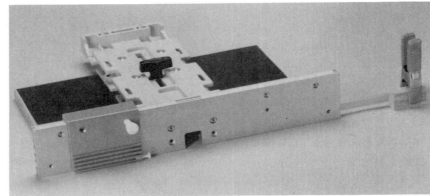

Illus. 28-6. The "original" Incra jig features an all-plastic adjustment mechanism and extruded aluminum fence, shown here with in-feed and out-feed stops.

you can remove the fence on these systems and attach your own (Illus. 28-7) and still utilize the systems' precise adjustment mechanisms.

The adjustment mechanisms differ considerably between the Incra and JoinTECH systems. The adjustment mechanism on the Incra original jig is essentially a two-piece plastic-moulded body (Illus. 28-8) with parts that have very fine saw-tooth-like surfaces ("racks") that interlock with each other. The teeth are spaced at ¹⁄₃₂″ and, consequently, when you clamp them together the opposing pairs wedge in precisely, eliminating any possible shifting as they are tightened together. Incra's Pro model is of solid, anodized aluminum construction with the same kind of mechanism.

The JoinTECH system is made of brass, steel, and aluminum components. Its adjustment comes from a precision-machined chrome lead screw with 32 threads per inch (Illus. 28-9). A push-button system releases the fence so it can be moved quickly to any position. It will also reengage an indexing block to the thread surface with a high degree of accuracy. The JoinTECH system

Illus. 28-7. The "original" Incra positioning jig attached to a wood fence.

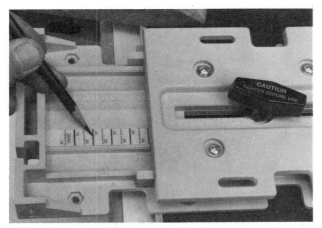

Illus. 28-8. A "template" on the lower body half of the Incra jig. This is an interchangeable strip of calibrated plastic that gives precise fence settings for the cutting pattern desired. Incra refers to them as templates.

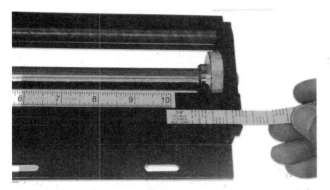

Illus. 28-9. Inserting a template into a shallow dovetailed slot on the JoinTECH jig. This view also clearly shows the finely threaded lead screw above with the vernier thumb wheel.

Illus. 28-10. The right-angle fixture on the Incra system. Note the hooked aluminum shape that fits over the fence.

has a built-in microadjustment in the form of a vernier thumb-wheel control calibrated in .002″ locking steps anywhere between the ⅟₃₂″ settings (Illus. 28-9).

An optional accessory for the Incra jig, called the Incra Mike, is a positioner with which you can set the fence anywhere between the ⅟₃₂″ drop-in settings, forward or rearwards, in .001″ steps.

Another structural difference between the two systems is the design and operation of their right-angle feed fixtures. The fixture on the Incra system (Illus. 28-10) is aluminum with an aluminum hook that straddles the top of the fence, and two plywood runners that ride on the router-table surface. Smoothness of operation improves when paste was is used on the router-table surface. The right-angle feed fixture on the JoinTECH system slides

more easily. It is made of slippery polyethylene with a male dovetail that fits a dovetail track in the top of the fence (Illus. 28-11 and 28-12). It has just one plastic runner that rides on the table. Both systems have a 6″ vertical surface for clamping the workpieces.

The fence stops on both systems are different, but well engineered and effective (Illus. 28-12–28-15). Stop extenders stored inside the fence pull out and enable you to use stops well beyond the length of the fence itself (Illus. 28-6).

So how do these precision fence systems actually make all those intriguing decorative joints without the need for measuring? All the settings required for routing a particular joint are preplanned for you. Each is basically preprogrammed and recorded with markings on printed plastic strips or "templates." When you insert or attach a plastic strip to the adjustment mechanism on either system, you align it with a hairline cursor (Illus. 28-16). When the adjustment mechanism is locked in at this setting, the cut is made at the precise fence position required.

Illus. 28-11. An operational setup on the JoinTECH system. Note the templates, one for in- and out-fence adjustment and one on the fence to set stops to control the length of cut.

Illus. 28-12. The fence stop on the Join-TECH system is made of polyethylene. Here the stop is being zeroed-out against the bit and the template in the fence is moved to where zero lines up with the near edge of the stop.

Illus. 28-13. The fence stop on the Incra system. Adjustments between the $1/32''$ settings can be fine-tuned with a nylon stop screw.

Illus. 28-14. The Incra fence stop, with a good view of the ⅟₃₂″ saw-tooth positioning rack.

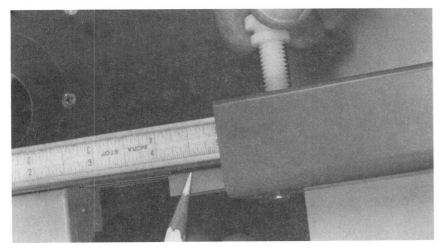

Illus. 28-15. Templates also are used on the top edge of the fence of the Incra jigs to set stops. Note the matching saw-tooth, ⅟₃₂″ pitch "racks" on the back of the fence that mesh with the "racks" on the stop.

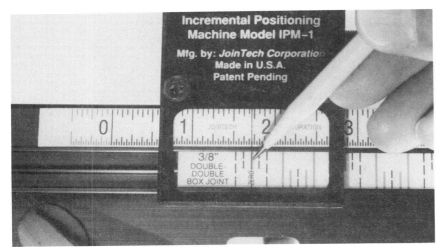

Illus. 28-16. A close-up look at the optical cursor on the JoinTECH system shows the fence adjustment on a calibrated "template."

Though mastering these systems is not difficult, it will take some trial and error. Incra and JoinTECH supply general instructions and full-size plans or cutting patterns. By placing your workpiece directly on these patterns, you can determine where the placement of cuts will be (Illus. 28-17).

Neither system has any provision for vacuum-chip collection, and neither system provides a guard that might be used for work that's fed flat over the table surface. Understandably, it's difficult to guard work when it is fed on end. These systems work best on router tables without mitre gauge slots. Some woodworkers may make tables primarily for these systems, with the router-mounting plates offset to one side or end.

One word of caution regarding the use of these jigs to widen slots, grooves, and dadoes: Be absolutely sure that you cut on the correct side, so the uncut stock is not trapped between the fence and the bit—a situation in which a very serious kickback can occur. Be sure to use push sticks or blocks and to be extremely careful when routing small, short pieces of stock, which seem to be cut often (see Chapter 22).

Remember that every cut of every joint requires that you make a fence adjustment. You must concentrate on the adjustment—the color coding of the templates for certain cuts helps, but somehow it is easy to make a mistake. With a Leigh jig, Keller jig, or Omnijig, once you have made the setup you complete all cuts concentrating on the cuts rather than adjustments. The latter jigs are also best used for routing joints on large pieces that are too thick, wide, or long to be handled on the router table.

The precision positioning systems are great for those individuals who love to experiment, who concentrate more on smaller projects or smaller components, and want to create some spectacular router work. Incidentally, Incra has an entire book—*Incra Jig Projects and Techniques*—devoted to a variety of intriguing projects, with complete plans for Incra jig owners.

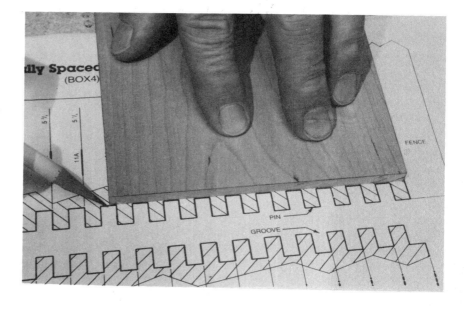

Illus. 28-17. You can determine how box-joint cuts will be spaced in the width of your workpiece by placing the workpiece over the manufacturer's full-size patterns.

29
Vacuum Air System

This chapter describes how to inexpensively and easily equip your shop with an exceptionally useful vacuum air system. A vacuum system allows you to hold flat pieces one against another without clamps or any kind of mechanical fasteners (Illus. 29-1). Turn it on and the vacuum sucks one part (bench clamp, template, or jig) against another flat surface (workbench, workpiece, or a machine worktable). Shut it off and the pieces are instantly released.

A vacuum air system can be used in dozens of ways. Some common ones include: (1) making templates and patterns of any size and any profile shape that attach to each other; (2) making a bench chuck (clamp) or vacuum table that will hold the workpiece flat without clamps to obstruct surface sanding, while you use a hand plane, biscuit joint, router, or other tool; (3) making self-gripping jigs and fixtures for machines or portable routing, such as a circle-cutting jig that requires no center hole to be drilled into the work surface, or a vacuum featherboard or vacuum stops; and (4) making vacuum-holding jigs for assembly work, to hold parts in place while you use both hands to drive home screws, etc.

Vacuum Air-System Components

Vacuum Pumps

Vacuum pumps essentially do the opposite of air compressors in that they pull in rather than push out the air. Vacuum pumps are available in many different types. Commercial vacuum pumps vary greatly in size and price. Vacuum pumps can also be salvaged from discarded household refrigerators (Illus. 29-2) and farm milking machines, as illustrated and described in my book *Router Jigs & Techniques*.

Refrigerator Pumps Refrigerator pumps (Illus. 29-2) are easy to find from discarded home refrigerators. Just remove the pump and disconnect the extra tubes. Plug it into an outlet, and find the intake. Do not be surprised if it does not have any more suction than that which is sucked through a soda straw. This is a sufficient amount.

Venturi Vacuum Pump You can also purchase an inexpensive venturi-type vacuum pump (Illus. 29-3–29-5). These types of pumps are also called vacuum generators. They require compressed air, but the compressor need

Illus. 29-1. A vacuum template in use. Note that the air hose in this case exits from the top; it can also be made to exit from the side if more convenient.

not be larger than a small ½ hp unit. These pumps generally work best at just 60–80 psi (pounds per square inch) rather than at the 100 psi or greater range. Though they are called pumps and generators, they do not have any moving parts or require any maintenance. They are very small (just 4″ to 6″ long).

It's interesting just how compressed air is used to create vacuum air. The compressed air is forced through a small orifice, which causes the air to speed up. It then exits through a specially designed funnel that allows the air to expand in a controlled manner, creating a lower pressure or vacuum.

You only need enough vacuum to evacuate all of the air out of a small cavity that you make with tape ⅛″ thick. Atmospheric pressure, approximately 15 pounds per square inch (at sea level), will exert a pulling or clamping force that's measured in inches of mercury (Hg). A perfect vacuum is 29.92 Hg. Because of material porosity, etc., it is realistic to expect a mercury level of less than 29.92 Hg. As a general rule, every 2″ of Hg will develop one psi, or 14.7 pounds of pressure at a perfect vacuum. This is essentially a 2:1 ratio. Thus, if, for example, your vacuum gauge reads 26″ of mercury, it translates to 13

pounds per square inch, or 1,872 pounds per square foot of surface area (26 ÷ 2 × 144). This is more than sufficient for any woodworking clamping or holding requirement.

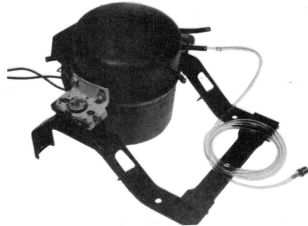

Illus. 29-2. This salvaged vacuum pump/motor from an ordinary discarded refrigerator is all you need to set up an effective home- or production-shop vacuum air clamping and template system.

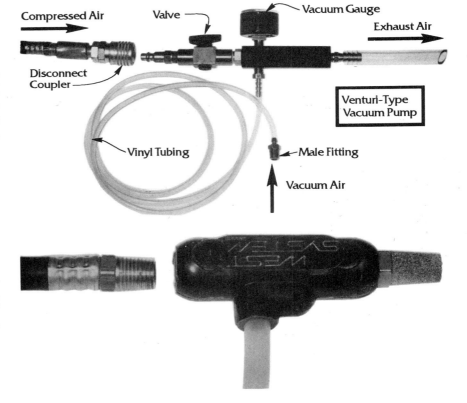

Illus. 29-3. The essential parts and air flow of a venturi-type vacuum pump. The pump shown here is of rugged, all-aluminum construction and is available from Quality Vakuum Products, Inc., Lincoln, Massachusetts.

Illus. 29-4. The vacuum generator available from Gougeon Bros. Inc., Bay City, Michigan, measures only 1¼″ × 4″ overall. It is of plastic construction, with a metal interior throat.

Illus. 29-5. The Vac-U-Clamp® kit from Vacuum Pressing Systems, Brunswick, Maine.

Muffler

Making Vacuum-Routing Templates and Clamping Fixtures

Any suitable nonporous material can be used, such as plastic, aluminum plate, sealed woods, and plywood material such as Baltic birch. The best vacuum-routing templates and clamping fixtures are made with plastic-laminate surfaces and sealed edges. However, Illus. 29-1 shows a template made of ½″ unsealed Baltic birch which works fine. For best results, use wood base material that is not thinner than ½″. However, plastic as thin as ¼″ and aluminum as thin as ⅜″ work fine.

All hoses and fittings needed can be purchased at the local hardware store (Illus. 29-6). Turn the pipe threaded end of a connector fitting (Illus. 29-7) into a hole drilled anywhere through the template's cavity area. Use the fitting as a self-tap. The fitting should fit tightly into the predrilled hole. To be sure it does, apply some silicone or epoxy glue around the fitting to ensure there's no air leakage.

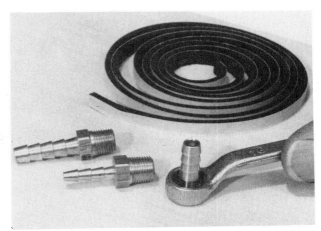

Illus. 29-7. Turn a connector fitting into a properly sized hole drilled through the template, so its threads will be self-tapping.

Vacuum Tape

Apply a closed-cell, thin tape that's ⅛″ to ¼″ thick and ¼″ to ⅜″ wide around the perimeter of the template, to create a clamping cavity (Illus. 29-8). You can butt-joint the tape, but be sure there are no visible voids. The best tape to use is a special vacuum tape purchased from a supplier that's ⅛″ thick and ⅜″ wide. If it's too thick, there will be some lateral movement between the template and workpiece. As a temporary measure, you can purchase

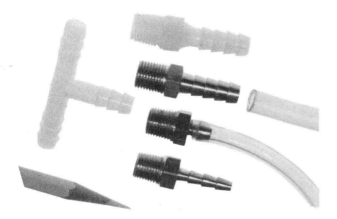

Illus. 29-6. Nylon and brass stem fittings. Note their male pipe threaded and serrated ends for inserting the fittings into flexible vinyl tubing.

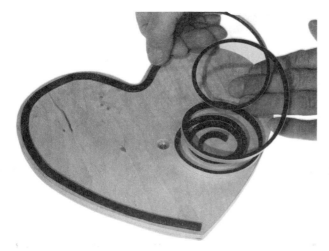

Illus. 29-8. Apply closed-cell foam tape around the perimeter of the template. Be sure to butt the ends tightly without any gaps. Note the air evacuation hole.

closed-cell vinyl foam-tape weatherstripping from your local hardware store, but it wears out and will lose its capacity to spring back if used continuously.

If either the template material or the workpiece is too thin or flexible, the vacuum may draw the parts so close together that they close the air escape hole and break the vacuum. To counteract this, place short strips of vacuum tape near the hole for support (Illus. 29-9).

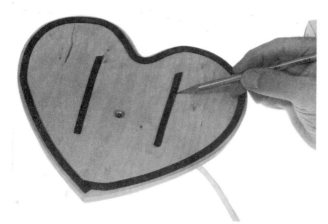

Illus. 29-9. Short strips placed alongside the hole ensure that the template will not be drawn too tightly against the work-piece. Thin material can bend or deflect. This closes the air escape hole, which breaks the vacuum.

Double-Sided Vacuum Clamp

A vacuum clamp can be of any size and shape. A vacuum bench-top clamp and the type of clamp used on overhead

pin routers *between* the workpiece and the template are similar and easy to make. Illus. 29-10 shows a rectangular vacuum clamp that you can use to hold workpieces and templates of any profile shape to each other. It is simply a block with vacuum cavities on both sides and of a thickness that allows the air-extraction connection to exit from the edge.

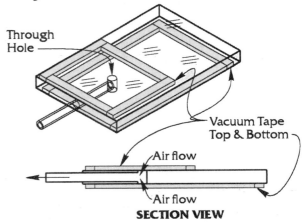

Illus. 29-10. This double-sided vacuum clamp will hold itself to a workbench with one side, and the workpiece on top of it with the other.

When used for hand-held routing, the vacuum clamp will hold itself to the workbench and hold the workpiece clamped on top of it when the pump is activated. Here is a design suggestion: Make two different-size cavities, one on each side of the vacuum clamp. This will allow you to hold different-size workpieces with one basic clamp. In either case, the workpiece will be held secure with the full surface and all edges ready and unobstructed for routing.

You may want to dedicate a table just for vacuum-clamping. Commercially made tables such as the one shown in Illus. 29-11 are available, but it's more fun and cost-efficient to make your own. You can also incorporate a lazy Susan into your design. See *Router Jigs & Techniques* for a plan. Later on, you may decide to add a foot control to keep both your hands free.

Vacuum Jigs and Fixtures

By now you should be able to visualize all the ways you can employ a vacuum system in your jig and fixture designs. Illus. 29-12 shows a circle-cutting jig with a vacuum-clamped centerpoint that will not damage or mark the work surface. It also shows my two-way feather-board that I use horizontally (on a table as a hold-in) or

Illus. 29-11. A 14″ × 14″ bench-top vacuum table will hold any size material from 4″ × 6″ to 4′ × 4′. It has its own pump that operates from compressed air. This unit is available from Unique Machine & Tool Co., Tempe, Arizona.

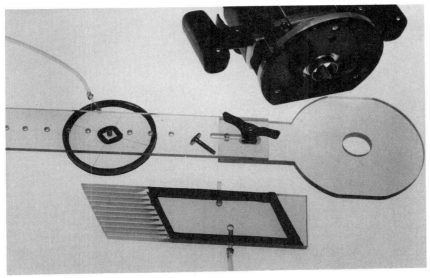

Illus. 29-12. The author's vacuum-center, adjustable circle-cutting jig and two-way vacuum featherboard made of polycarbonate plastic. Note the concentric rings of vacuum tape applied to the downside of the pivot pin vacuum chuck of the circle-cutting jig.

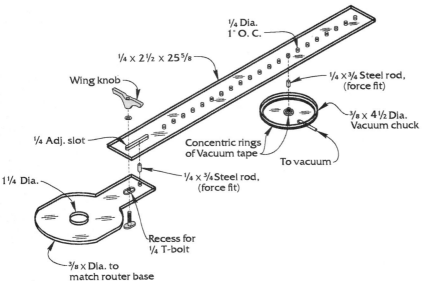

Illus. 29-13. Details for making the vacuum-center, adjustable circle-cutting jig. As designed, this one is linked to the router with a Porter-Cable No. 40 42021 template guide (cut shorter), which permits the use of bits up to 1″ in diameter.

vertically (on the high fence as a hold-down). Note that it has two individual out-of-the-edge air-escape holes (one from each surface), and that vacuum tape is applied to both surfaces.

Illus. 29-13 gives the essential details for making the fully adjustable circle-cutting guide. The design incorporates a short-range sliding adjustment that is used in conjunction with the pin and hole increments to produce any radius desired without marking or putting a hole in the work surface.

30
Overarm Routers

Routers held or supported vertically over a router table are known as overarm routers. They are also called pin routers because most have a table pin (sometimes retracting) in the table that is used to guide patterns and templates for duplication work. Pin routers can be used to make many different router cuts, including practically everything with the exception of three-dimensional carvings. Some have mechanical and air-controlled router plunging and retraction capabilities that permit safe bit entry when cutting interiors and inside profiles.

These routers, and the systems used to control them, are generally very expensive, heavy, industrial-production machines. However, a few scaled-down versions (Illus. 30-1–30-3) are available that are geared towards the home-craft and light-production woodworking markets.

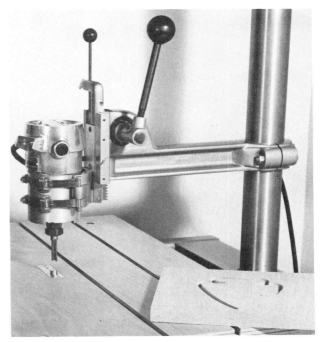

Illus. 30-2. The major components of the Shopsmith overarm mechanism shown without guard and dust-extraction hookup. Note the provision for vertical travel for mounting the router, and a twist-lock vertical-feed lever with a rack-and-pinion slide device.

Shop-Made Arm Attachments

A shop-made arm attachment (Illus. 30-4–30-8) can be made to convert the drill press into a light-production overarm router. A very simple accessory can also be custom-made to combine the high speed of the router with all the actions of the radial arm saw, which in itself is very versatile (Illus. 30-10–30-14). To take each of the best features each of these power tools has to offer, remove the saw blade and make (or have made) a similar router-mounting device.

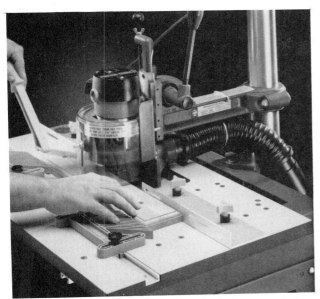

Illus. 30-1. Shopsmith's router system holds the router in two different positions—under the table or above it on an overarm, as shown here.

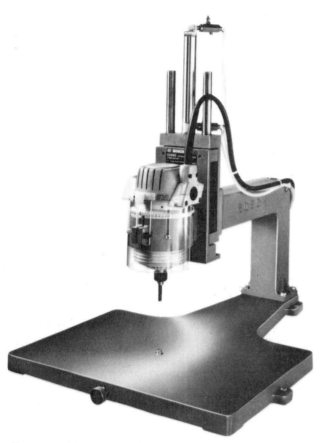

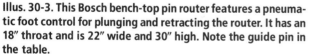

Illus. 30-3. This Bosch bench-top pin router features a pneumatic foot control for plunging and retracting the router. It has an 18″ throat and is 22″ wide and 30″ high. Note the guide pin in the table.

Illus. 30-4. This shop-made fixed overarm routing accessory for your drill press is easy to make, but it does not have plunge-entry capabilities.

Illus. 30-5. The shop-made rigid over-arm column with a hose clamp and a bolt gripped in the drill-press chuck. The quill is locked to prevent vertical travel.

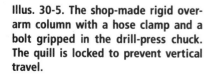

Illus. 30-6 (left). Drill the bolt hole with the arm clamped tightly to the column, as shown. Illus. 30-7 (above). A rabbeting setup with a shop-made plastic ring guard which can and should be used whenever possible.

Illus. 30-8. Simple grooving with the shop-made, drill-press overarm-routing fixture.

Illus. 30-9. Fence-guided mortising and grooving with the Shopsmith overarm, which features a plunge-entry mechanism. The workpiece is fed from left to right, against the bit rotation.

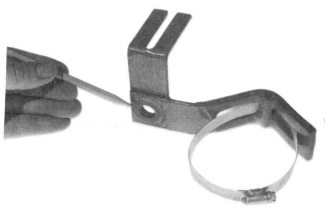

Illus. 30-10. This simple routing jig (made of metal) slips onto the saw arbor of the radial arm saw.

Illus. 30-11. The bracket mounts onto the saw arbor like a saw blade, and the guard stud on the motor housing keeps it rigid.

Illus. 30-12. The router mounted to the arbor of a radial arm saw.

Illus. 30-13. The radial-arm-saw routing attachment in use for indexed grooving to make decorative moulding.

Illus. 30-14. A stop clamp on the saw's arm track and a work-holding fixture assisted in making stopped slot cuts for louvres in these door stiles.

The radial-arm-saw router mount is simply made from 1/8" flat steel with a bored hole so it slips over the arbor of the radial-arm-saw motor. It is kept rigid with a secondary anchor to the blade-guard pin, found on the motor housing of most radial saws. A slot in the bent area of the arm permits a large hose clamp to be inserted, to secure the router vertically. Make a stiff-cardboard mock-up of the router mount to ensure that it will fit the saw. Then take the mock-up to a metal shop and have a router mount made. You may have to slightly bend or twist the router mount in a vise, with leverage from an adjustable wrench, to get the router in a true vertical position. The one shown in the photographs in this section was dipped in liquid plastic to protect the router motor housing. Think of all the different kinds of cuts that are possible with this simple accessory, using it either fixed or in the horizontal travel mode.

Basic Overarm and Pin-Routing Applications

Basic overarm and pin-routing applications include the following operating techniques:

1. Straight-line work (Illus. 30-7–30-9)
2. Edge-forming with piloted bits (Illus. 30-15)
3. Pin-routing of outside contours with and without templates or patterns (Illus. 30-16 and 30-17)
4. Forming internal openings and piercing with patterns (Illus. 30-18–30-20)
5. Freehand routing (Illus. 30-21 and 30-22)

Basic overarm and pin-routing techniques can create an endless variety of different profiles and shapes. Overarm and pin-routing have some distinct advantages over

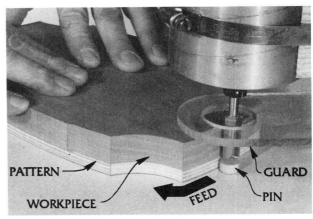

Illus. 30-17. Pin-routing with a pattern using the drill press overarm routing fixture.

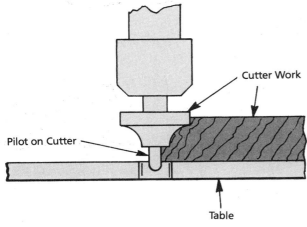

Illus. 30-15. Edge-forming with a piloted bit.

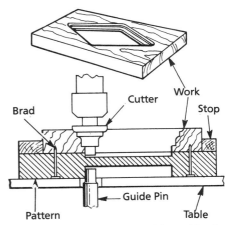

Illus. 30-18. Pattern/template-routing with a table pin to produce an inside opening.

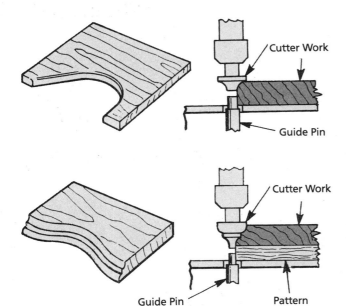

Illus. 30-16. Above: pin-routing without a template pattern. Below: contoured edge-forming with a template pattern.

Illus. 30-19. Tracing grooves in an attached pattern over a pin creates surface designs and/or cuts inside openings. (Photo courtesy of Shopsmith)

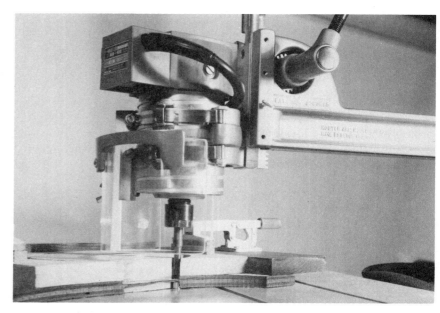

Illus. 30-20. A close-up showing a cutaway of the pattern and workpiece revealing the pin groove of the pattern and the reproduced cut made by the bit. Note that the pattern is "boxed," that is, it cradles the workpiece, which is also clamped securely to the pattern.

Illus. 30-21. Raised-letter sign work as done here involves fence-guided work for the long straight-border cuts and freehand feeding.

Illus. 30-22. Freehand levelling of a wood slab.

conventional table-routing operations discussed in previous chapters. The essential advantage is that, unlike table routing, the routing is done with the operator's full view of the cutting action. This allows stopped grooving, mortising (Illus. 30-9), dadoing, and other similar jobs to be done much faster with less setup time and inconvenience involved. Another important advantage is that with the cutting done above the work, it simplifies some machine template-routing techniques, which can be problematic on router tables.

With pin and overarm work, the template is fastened under the blank workpiece and the operator can easily see where the cutting is taking place and when the job is completed. Overall, using fixtures in pin and overarm work is a lot easier too, although not for all jobs.

There are some distinct disadvantages to overarm-routing too. First, the entire bit is above the table, dangerously fully exposed and difficult to guard properly for many operations. Second, a dust-extraction system is more difficult to rig up when you use overarm routers. Third, any chips getting under the work will raise it, making cuts deeper than desired. Fourth, although hand-feeding must obviously always be done very carefully (against the rotation), you must be particularly careful in certain situations such as when edge-routing into sharp inside corners. The bit has a tendency to grab and reject the workpiece.

Industry makes very good use of overarm and pin-routing machines. Many are computer-controlled, with a variety of automatic accessory devices, and even computerized feeders. Space just does not allow me to discuss or illustrate all these important and interesting high-production processes.

Obviously, all the capabilities of overarm and pin-routing devices and machines have not been illustrated in this chapter. Some can also be used to drill holes, machine mouldings, raise panels, and cut recesses of all types, such as those used to hollow out boxes and trays. They can be set up with a planing fence to function as a vertical jointer or full-edge shaper and can handle many other operations, including trimming plastic laminates and making delicate inlays.

Anyone getting into a production operation will find that it is a good idea to rig up a foot-operated switch connected to the router motor. Thus, you can hold the work at all times with both hands and maintain control, should it be necessary to quickly shut the router off. It's also important to support long and big workpieces with table extensions, a roller stand, etc., so the work does not

tip off the table; this can result in a cut deeper than desired—or worse, a kickback.

Radial-Arm Routing Machine

Illus. 30-23 and 30-24 show an unusual radial-arm routing machine produced by the Edge Finisher Corporation of Westport, Connecticut. It is designed for cutting and reproducing all types of shapes. It features a rotating table with a foot-controlled vacuum template system on a movable lower arm that permits feeding the work into ball-bearing-guided pattern and trimming bits.

Illus. 30-23. This radial-arm routing machine is designed for following templates and patterns to cut out contoured shapes. The lower arm moves horizontally; the table with a vacuum-clamp attachment holds and rotates the workpiece into the bit. (Photo courtesy of Edge Finisher Corp.)

Illus. 30-24. Vacuum-template-routing oval-shaped plastic parts on the radial-arm routing machine. Note the hand wheel for rotating the table, and that the work is rotated clockwise into the bit.

31
Special-Purpose Routing Devices

There are a growing number of router accessory devices or machines designed to do a narrow range of special cutting jobs. This chapter provides an overview of some of those products currently available that may have some possibilities for secondary uses.

Sears Mill-Works Accessory

The Sears Mill-Works accessory (Illus. 31-1 and 31-2) is essentially designed for making decorative cuts and dentil mouldings. It is of all-plastic construction and consists of a carriage and subframe. The router is mounted to the carriage, which, with a system of stops and indexing pins, locates and directs each cut. An optional tip-out sub-base can be used to make tapered or inclined cuts (Illus. 31-3–31-5).

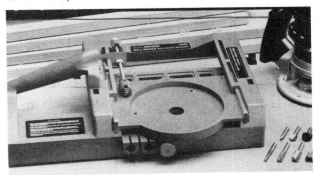

Illus. 31-2. A view showing the carriage to which the router is mounted. Note the adjustment for the length of cut to the left.

Illus. 31-1. The Sears Mill-Works accessory and some examples of the equally spaced cuts and dentil mouldings made with it.

Illus. 31-3. A bottom view of the Sears Mill-Works accessory. Note the bit in relation to an indexing pin used to space the cuts uniformly.

Illus. 31-4. An adjustable sub-base pivots from the frame for making taper or inclined cuts.

Illus. 31-5. The Sears Mill-Works attachment in use.

Hoffman Dovetail Joining Machines

The Hoffman dovetail joining machines are several router-driven machines that cut mating blind dovetail slots across joints of all kinds (Illus. 31-6). They are most commonly used to join, align, and strengthen flat mitred frames, etc., but can be applied to many fastening and assembly jobs. The system revolves around plastic butterfly-shaped dovetail "keys" with rippled surfaces that bite and pull the joint together when they are driven into the special dovetail slots made with these machines. Contact Hoffman Machine Co., Inc., of Centereach, New York, for more information.

Pinske Drainboard Machine Router

The Pinske drainboard machine router (Illus. 31-7) is designed to rout gently inclined drainboard grooved patterns directly into hard-surface countertops so water can run off into the kitchen sink. Various acrylic pattern designs are available for countertop use. It appears that this machine could also be used to duplicate other surface-cut designs and patterns into other materials as well, once a master pattern is created. Contact the Pinske Edge, Plato, Minnesota.

Fox Chair Seat Disher

The Fox chair seat disher (Illus. 31-8) is a C-frame double-arm machine that has a router mounted to its upper arm and a guide roller to its lower arm that follows the surface of the master pattern. The pattern follower and router move in unison on a shaft that slides from

Illus. 31-6. The Hoffman dovetail joining machines are used to rout stopped or blind dovetail slots into both profiles of the joint. Plastic keys that align and pull the joint tight are inserted into these slots.

front to back and side to side. The bits are special-ground, and are 1½″ in diameter.

Manual and automatic chair-seat dishing machines capable of carving up to 40 parts per hour with optional accessories are available from Fox Mechanical Products of New Holland, Pennsylvania.

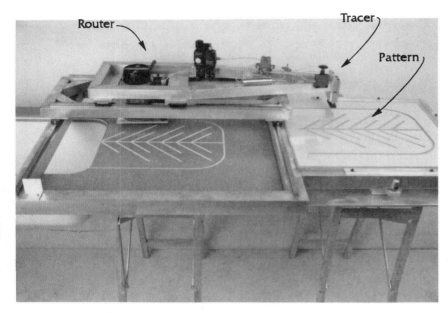

Illus. 31-7. The Pinske drainboard machine router reproduces grooved patterns into solid-surface countertops to direct water runoff into the kitchen sink.

Illus. 31-8. The Fox manual chair-seat disher.

32
Joint-Making and Multipurpose Routing Machines

A variety of router-based machines capable of generating a multitude of different joints are available for the serious router crafter. This chapter presents an overview of some of these devices and also takes a look at inverted routing machines, which are also becoming increasingly popular in small production shops.

Shopsmith Joint-Matic

The Shopsmith's Joint-Matic (Illus. 32-1 and 32-2) is a router-operated joint-making machine. The router is mounted horizontally on a crank-operated slide that

Illus. 32-2. A close look at the slide and router support, the adjusting crank, the scale, and table — the key components of the Shopsmith Joint-Matic.

moves up and down. The slide with router moves precisely $\frac{1}{16}''$ with each full turn. This vertical movement and precision control permits fine adjustments ($\frac{1}{4}$ turn $= \frac{1}{64}''$) and allows the user to make specialized joints with minimum setup (Illus. 32-3 and 32-4).

The worktable accommodates a mitre gauge, and the unit can be used with a number of optional accessories including an auxiliary table, bevel mitre gauge, featherboards, and push blocks. It weighs 65 pounds and measures $21\frac{1}{2}''$ wide $\times$ $33''$ long $\times$ $57''$ high.

Illus. 32-1. Shopsmith's Joint-Matic is a horizontal, router-operated machine.

Illus. 32-3. Box-joint routing on the Joint-Matic machine.

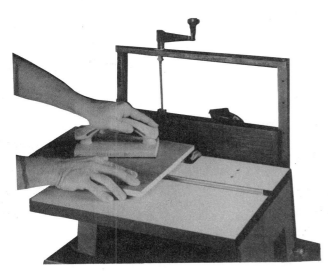

Illus. 32-4. Raised panel work with Shopsmith's auxiliary table accessory.

Illus. 32-5. The RBI Router Shop features a vertically adjustable table and a tilting router-mounting plate that permits the router to be used in a horizontal position, as shown, or in a conventional under-the-table mode.

Illus. 32-6. Joint work on RBI's Router Shop with the router-mounting plate vertical for horizontal routing.

RBI Router Shop

The RBI router shop (Illus. 32-5) is similar in function to the Shopsmith Joint-Matic, but it operates on a somewhat different principle. The worktable adjusts up and down in relation to a fixed router. It moves 1/16" with each turn of a crank and it has a full 12" travel range. This product not only has the provision for using the router horizontally (Illus. 32-6), but the router mounting plate is adjustable at 0- to 90-degree angles to the table, with positive stops at 22½, 30, 45, and 90 degrees. When it is folded down completely, you have a conventional router table (Illus. 32-7). Other accessories and a smaller version called the Router Shop II (which is actually more comparable to Shopsmith's Joint-Matic machine) are available from RB Industries, Harrisonville, Missouri.

Trend Routerack

The Trend Routerack (Illus. 32-8 and 32-9) is a universal router stand that allows hand-held routers to be mounted in four ways. Basically, it consists of 2″ steel pipe pillars and various mounting brackets of aluminum. When you use various accessories with the Routerack, you have a very versatile routing system capable of handling a wide variety of routing jobs (Illus. 32-10 and 32-11). The Routerack system is available from Trend Machinery Ltd., Watford, Hertfordshire, England.

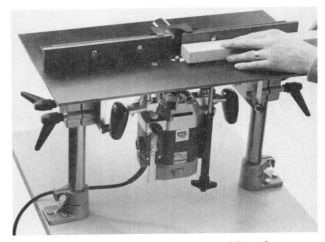

Illus. 32-9. The Routerack with a worktop and fence for conventional router-table operations.

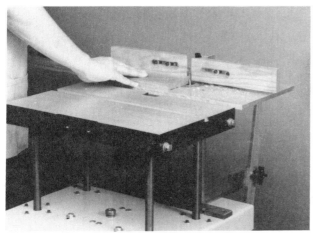

Illus. 32-7. When the mounting plate on the RBI Router Shop is folded down flat, the unit becomes a conventional router table with a vertical bit operation.

Illus. 32-10. The Trend Routerack with the router mounted in the horizontal mode for tenoning, rabbeting, etc.

Illus. 32-8. The Trend Routerack, shown with the router mounted in an overhead position, consists essentially of 2″ steel pipe pillars and various mounting brackets of aluminum. Here it is shown with a compound machining table accessory.

Illus. 32-11. Circle-routing with the Trend Routerack.

Multi-Router

The Multi-Router (Illus. 32-12), like many other machines, has two working surfaces that are perpendicular to one another. The vertical surface houses the router. The workpiece is held against the horizontal table, which also has a tilting capability for making angular cuts. This machine also incorporates a system of templates that are traced by a ball-bearing stylus. The router is counterbalanced, so it can be moved up and down with ease. Round and oblong tenons and double mortise-and-tenon joints (Illus. 32-13–32-15) are easy to make because all you have to do is follow the pattern of the template with the stylus. This machine with templates for box joints, dovetails, and different-size tenons is available from the JDS Co., Columbia, South Carolina.

Illus. 32-14. With the tenons cut, they are used to set the stops for making the mortise cuts. (Photo courtesy of Curtis Whittington)

Illus. 32-12. The Multi-Router.

Illus. 32-15. Completed double mortises cut on the Multi-Router by Curtis Whittington.

The Matchmaker

The Matchmaker (Illus. 32-16) is a product of the Woodworkers Supply Co., of Albuquerque, New Mexico. This is also a horizontally mounted router and essentially a template-controlled machine (Illus. 32-17 and 32-18). It has a work-holding table assembly that moves in two horizontal directions; the router attached to a stylus-

Illus. 32-13. A double mortise-and-tenon joint made on the Multi-Router. (Photo courtesy of Curtis Whittington)

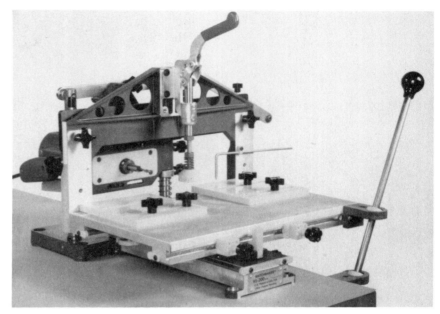

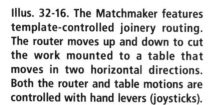

Illus. 32-16. The Matchmaker features template-controlled joinery routing. The router moves up and down to cut the work mounted to a table that moves in two horizontal directions. Both the router and table motions are controlled with hand levers (joysticks).

Illus. 32-17 (above). Here you can clearly see a stylus following a shutter mortise template. Illus. 32-18 (right). The setup for cutting the pins of a through-dovetail corner joint. Note the tapered templates, the stylus, and the router bit orientation of this Matchmaker routing machine.

guided mechanism moves in a third direction. One hand is used to control the direction of the router and the other to simultaneously control the router table movement. The stylus follows the pattern of the template and, as this is being done, the router cuts the shape of the template into the workpiece clamped to the table (Illus. 32-19). A variety of templates and an air work clamp with a foot control are also available.

The WoodRat

The WoodRat (Illus. 32-20) is a very different joint-making routing machine. It is essentially a wall-hung

Illus. 32-19 (above left). Tenon-cutting. Note the tenon template and stylus above the work and the router bit cutting below the work. Illus. 32-20 (above right). The WoodRat is a wall-hung routing machine designed for joinery and cabinetmaking. Note the plastic parallelogram in the upper left of the photograph which is used for the layout and spacing of joint cuts.

routing arrangement invented by British woodworker Martin Godfrey, for advanced joinery in furniture and cabinetmaking (Illus. 32-21–32-23). This machine consists of a large horizontal extruded-aluminum body or base. Inside this is another extrusion which functions as a work-holding carriage that can move the wood under a fixed router with a hand-wheel feed or can be positioned appropriately so the workpiece can be cut by moving the router rather than the wood. A short crosscutting action is used in which the router (mounted to a cross plate) slides on top of a base plate mounted to the horizontal aluminum extrusion base. The cross-feed cuts can be at 90 degrees or angled to do jobs such as cutting the sides of dovetail pins, etc.

The combination of router movement and/or work-feed movement provides great flexibility to produce nu-

Illus. 32-21. Various joints cut with the WoodRat.

Illus. 32-22. A variety of tenons made with the WoodRat.

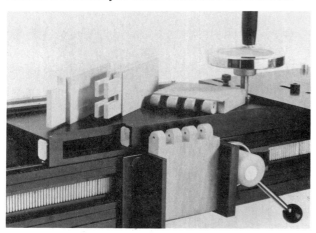

Illus. 32-23. Precision joints are cut without the use of templates. Note the eccentric cam-action clamp that holds the workpiece to the carriage face.

merous controlled joinery cuts. One distinct advantage of this unit is its ability to machine the ends of boards which otherwise are difficult to handle on end over a conventional router table or would require many templates if the same variety of cuts were made horizontally, as done on most other devices. The WoodRat offers an endless array of wood-jointing capabilities; however, it takes some time, practice, and experimentation before one can completely capitalize on its full range of joint-making potentials.

Stanfield Joiner/Shaper

The Stanfield Joiner/Shaper (Illus. 32-24 and 32-25) is another machine that features a horizontal router. This one is mounted to an air-cushioned carriage mechanism that moves and is adjustable in three directions on bearing guides: up and down (on air), laterally, and in and out. The router movement can be fixed in any or all of three travel axes, or stops can be set to control or limit the travel in any of the three directions. The router moves freely with hand-touch control, especially up and down. It is almost weightless because of the air-cylinder cushion supplied from a tank under the table. An abundant choice of templates for mortises, tenons, various dovetails (through, half blind, etc.), and box joints in many sizes is available, as is a fence. As the templates are followed with a stylus, the router is directed to cut the workpieces clamped to the table. You can also make your own templates for various shaping and cutting operations such as square, tapered table legs, etc. There's great potential for shaping and forming.

The fence accessory converts the router to a typical horizontal router table, and operations such as panel-

Illus. 32-24. The Stanfield joiner/shaper provides router movements in three axes: 23" laterally, 4" vertically, and 4½" in and out.

Illus. 32-25. Template-controlled tenon-cutting on the Stanfield joiner/shaper.

raising are done with the router fixed (clamped) so it does not move in any of its three travel axes. For more information, contact Stanfield Mfg., Inc., Albuquerque, New Mexico.

Inverted Pin Routers

Inverted pin routers (Illus. 32-26–32-28) are primarily designed for pattern- and template-routing. They are like conventional router tables except they have upcut plunge-router capabilities and interchangeable overarm pins that may or may not lower and retract automatically.

Hartlauer Duz Awl Machines

The Hartlauer Duz Awl machines (Illus. 32-26), made in Eugene, Oregon, are primarily sold to the sign industry, but they can be put to good use in general woodworking shops. They are made in two sizes, one with a 20″ throat and the other with a 36″ throat. The smaller machine is designed for a 1½ hp router motor which travels up and down on a ball-bearing movement mechanism. The larger machine has a 3 hp, multiple-speed router. Depressing the foot pedal retracts the bit below the table and simultaneously raises the over-head pin.

Onsrud and Delta Inverted Pin Routers

The Onsrud and Delta inverted pin routers (Illus. 32-27 and 32-28) are essentially the same machines. Both are made by C. R. Onsrud, Inc., of Troutman, North Caro-

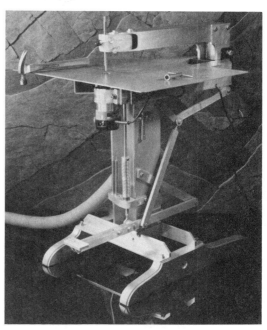

Illus. 32-26. The Hartlauer Duz-Awl Junior pin router has a 20″ throat and a vertical router travel controlled by a foot pedal.

Illus. 32-27. This Onsrud inverted pin router is just one of many models produced by the company.

Illus. 32-28. The Delta version of the Onsrud pin router is made by Onsrud.

lina. These are versatile machines, even though their best function is duplication from patterns and templates (Illus. 32-29). They can obviously be fitted with a fence for straight-line work.

The setup for template-routing is shown in Illus. 32-30 and 32-31. A 3 hp Porter-Cable router motor unit supported below the table surface in a sound-insulated cabinet provides the power (Illus. 32-32). The cutter height and plunge-router motion is controlled mechanically by a foot pedal. The same pedal automatically extends and retracts the guide pin of the overarm by means of a pneumatic air cylinder. The throat capacity of the Onsrud and Delta inverted pin routers is 20″, and both machines have an acrylic-topped table surface that is 24″ wide × 36″ long. The machine also features a

well-designed chip-extraction arrangement (Illus. 32-31) and has a net weight of approximately 260 pounds.

Illus. 32-29. Some examples of products produced with inverted routers.

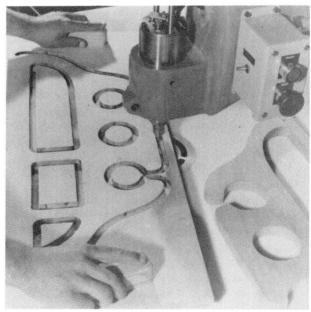

Illus. 32-30. Inverted routing exposes the operator to less risk than overarm (overhead) routing because the bit is either below the table or cutting into the workpiece.

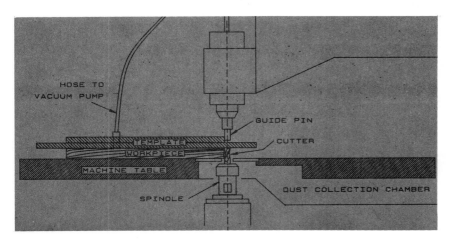

Illus. 32-31. This cutaway drawing of an Onsrud inverted router illustrates the workpiece sandwiched between the pattern and the machine table. Note the efficient chip-and-dust-removal system and the simple design of the vacuum template-holding system.

Illus. 32-32. An under-the-table view of the router with its up-plunge fixturing and the acoustical insulation to dampen noise levels. Here the operator adjusts a depth-of-cut stop.

33
Router Turning Machines

A machine that combines the stock-holding and rotation capabilities of a spindle lathe with a controlled router feed can produce astonishing wood products. This chapter examines two products designed so the hobbyist and creative woodworker can simulate the work done on expensive production copy lathes.

Sears Router Crafter

The Sears Router Crafter (Illus. 33-1) is a light-duty machine that weighs just 11½ pounds. However, when you use it with your own router, the Router Crafter allows you to do much more than is possible with an ordinary spindle-turning lathe. You can make distinctive turnings of all kinds, including candlesticks and lamp and table legs with beads, flutes, spirals, ropes or spirals, and twists (either right or left) in parallel or tapered configurations (Illus. 33-2). Almost any conceivable design can be reproduced exactly, time after time. Four table legs can be made, each exactly like the other. The workpiece-size capacity of the Router Crafter is approximately 3″ square by 36″ in length.

There are four basic types of operational techniques that can be performed on the Sears Router Crafter. These techniques used in combination with one another will result in hundreds of design configurations (Illus. 33-4). The techniques are:

1. *Straight beading and fluting.* This includes cuts made lengthwise or parallel to the workpiece. They may be of a straight, tapered, or contoured style.
2. *"Roping" or spiralling*—both right- and left-hand. These cuts can also be made on straight or tapered surfaces.
3. *Making beads, coves, and flats around the surface of the workpiece.* These cuts are determined by the cutting shapes of the router bit used.
4. *Contour template-turning*, in which a router follows the edge of a flat template attached to the front of the machine. This operational feature has some definite limitations because the machine *does not* copy in a 1-to-1 ratio.

The Router Crafter is not a heavy-duty production

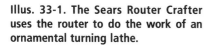

Illus. 33-1. The Sears Router Crafter uses the router to do the work of an ornamental turning lathe.

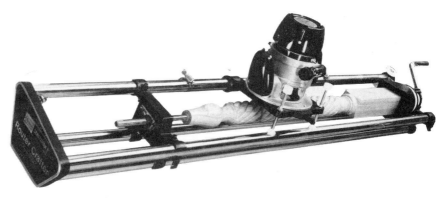

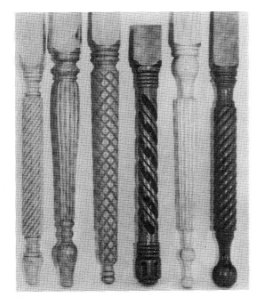

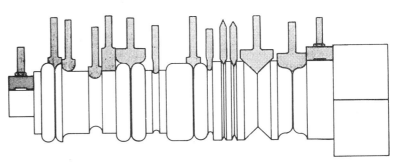

Illus. 33-2 (left). These furniture legs were made with a Router Crafter. Illus. 33-3 (above). Bits and turning cuts.

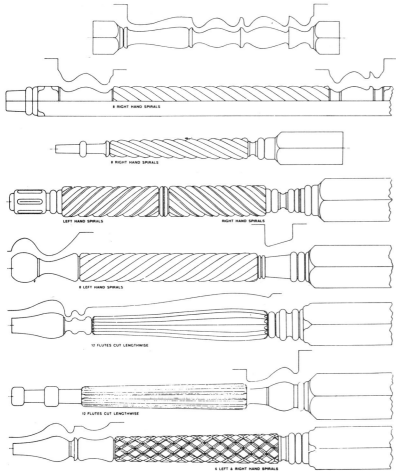

Illus. 33-4. Some of the turning configurations that can be produced with the Router Crafter. Those with lines next to the profile indicate the need for templates and their relative shapes for duplication.

machine. This accessory will provide hours of fun and entertainment for the hobbyist woodworker. Several successive shallower passes are required to complete certain cuts. However, it is possible to make some cuts ⅛″ to 3/16″ deep on softwoods in one pass; this actually reduces the diameter by twice that amount, so for some jobs the final shape can be arrived at fairly quickly.

The Router Crafter consists of a headstock and a tail stock connected by four steel tubes. A carriage to which a router is mounted travels along these tubes. The headstock incorporates a system of cables (used for spiralling work), an indexing mechanism (used for straight-line work, fluting, etc.), and a hand crank. When the crank is engaged, it works the drum-and-cable connections and synchronizes the carriage-travel-to-workpiece-rotation for spiralling work. The carriage is only connected to the cable for spiral cuts. Illus. 33-5–33-12 shows some of the cutting techniques in progress.

Illus. 33-5. To convert square stock to round stock, first make a series of successive passes on the stock while it is being held stationary. Initial round roughing cuts results in a series of 24 "flats" completely around the workpiece.

Illus. 33-6. To remove the "flats," simultaneously move the router while rotating the workpiece.

Illus. 33-7. A shop-made template is required to duplicate spindle turnings of a specific contour. Note that the profile shape of the template is exaggerated, and not identical to the contoured profile of the finished turning shown. The pencil points to the template follower of the carriage, which rides along the edge of the template. This controls the up-and-down feed of the router.

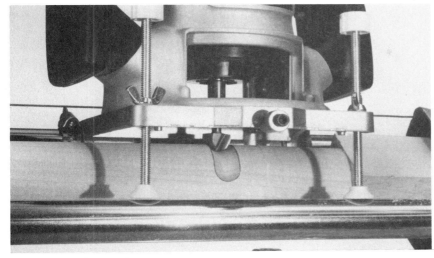

Illus. 33-8. Turning a cove around a workpiece.

Illus. 33-9. A cove bit cuts equally spaced flutes lengthwise on this tapered table leg.

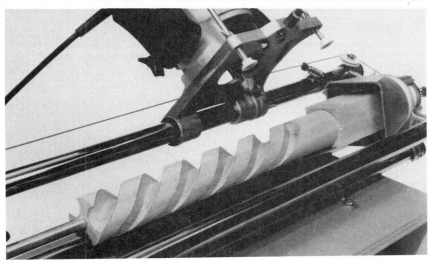

Illus. 33-10. Spiral cuts made with a core-box bit.

Illus. 33-11. A close-up view of spiralling using an ogee point-cutting bit.

Illus. 33-12. Diamond or "pineapple" effects are achieved by employing both right- and left-hand spiralling techniques.

Woodchuck Indexable Milling System

The Woodchuck Indexable Milling System (Illus. 33-13) is a lathe-like machine with indexing and tapering facilities that is used with the router and bits to produce spindles of various decorative styles. Typical uses revolve around three basic areas of work: (1) sizing and truing stock; (2) indexed joinery such as slotted dovetails, mortises, etc.; and (3) multisided cross sections or decorative cuts that include fluting, beading, rope moulding, spiral twists, circular shaping, twisted flutes, and various combinations of these cuts.

The Woodchuck Indexable Milling System can be used on a workpiece as great as 12″ in diameter and 80″ long. There are two models available with net weights of approximately 75 pounds and 105 pounds. They are made essentially of stamped and machined metal. A linear-drive screw and a laser-cut spur-gear reduction set are used for controlled spiralling operations. The operational concept is that the workpiece mounts between centers similar to those on a lathe, and the Woodchuck provides cutting control on five independent axes. The router with the aid of a table is positioned or moved to cut in an X, Y, or Z axis. A plunge router works great, but fixed base routers will also work. The workpiece can be indexed (fixed in any rotational position) and milled with the router on any number of surfaces with the use of a variety of indexing plates (Illus. 33-14). Independent controls adjust each end of the worktable; this permits the milling of tapers. The spiral drive permits right- and left-hand spiral-cutting (Illus. 33-15 and 33-16).

The Woodchuck Indexable Milling System is available direct from the manufacturer, Phantom Engineering, Inc., of Provo, Utah.

Illus. 33-13. The Woodchuck Indexable Milling System is a lathe-like machine with various mechanical work-holding and driving features that is used with a router.

Illus. 33-14. A view showing the 12-position indexing head and the crank for the linear-drive screw.

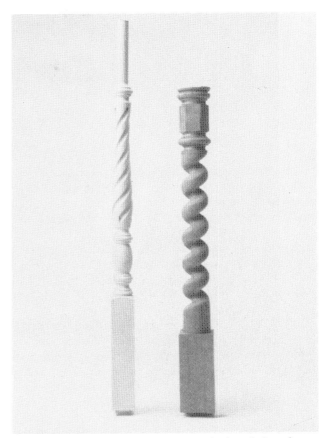

Illus. 33-15. A close-up look at some single spiral work produced with the Woodchuck Router Milling System, one of two models available from Phantom Engineering, Provo, Utah.

Illus. 33-16. Two projects made with the Woodchuck.

34
Router Carving Machines

Production carving machines are very fascinating pieces of equipment which have been around since the 1850s. Only in recent years, however, have a number of carving machines become available for home shops and serious woodworkers. Most carry a high-speed router motor as the power source or carving head. These are generally light-duty, inexpensive machines compared to those used in high-production factories for making multiple carvings in large quantities of table legs, gun stocks, and similar items.

It is not even necessary to be artistic or creative to make carvings with these machines. Most carving machines today are used primarily to copy or reproduce from master patterns. These machines are very easy to set up and use; they require no special skills. The crucial requirement is patience—even machine-carving takes time. However, it is still much faster and more accurate to machine-carve an object by tracing a pattern or another carving than it is to do it all by hand, from scratch. Many experienced hand carvers use carving machines to duplicate their originals.

Most machines have the ability to reproduce very fine details. Some have greater carving-size capacities, more reach, and overall convenience than others. The quality of the carvings seems to improve as the purchase price of the machine increases, but this is not always the case.

Anyone who intends to carve for fun or for profit should consider the various types of router carving machines described in this chapter. Some of the machines are simply versions of router pantographs. Not all are capable of reproducing or copying fully rounded contours of true three-dimensional shapes. Some only do two-dimensional work, such as flat wood signs.

Kimball Carving Machines

The Kimball carving machines (Illus. 34-1 and 34-2) are essentially of aluminum construction and designed for carving wood signs. Made in Providence, Rhode Island, by the Kimball Machine Co., these machines are avail-

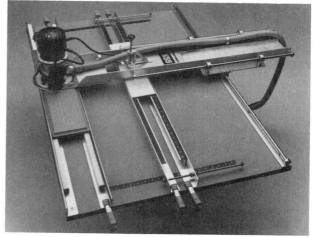

Illus. 34-1. The Kimball Machine Company's model K55 middle-of-the-line sign-carving machine. Note the vacuum dust-collection accessory.

Illus. 34-2. The Kimball Machine Company's model K90 top-of-the-line carving machine can carve signs 56″ wide and of an unlimited length.

Illus. 34-3. Samples of the various templates used with the Kimball sign-carving machines.

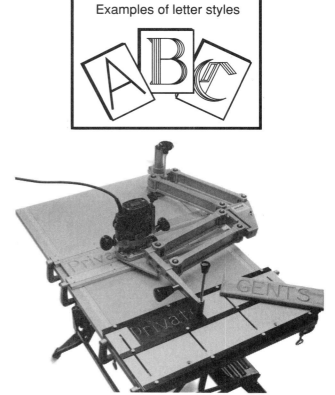

Illus. 34-4. The Trend sign router copies various templates on a one-to-one ratio basis, as shown. Note the examples of template designs shown above.

able in three different models and sign-size capacities. The largest (Illus. 34-2) will take stock of any length and up to 56″ wide. These machines are template-driven (Illus. 34-3), copy on a one-to-one size basis, and feature a spring-loaded stylus and a "floating carriage" system in an X- and Y-movement axis. The carriage components are ball-bearing bushings and pillow blocks mounted on high-carbon steel shafts that ensure a smooth, wobble-free motion. The carving power unit is a hand-held portable router mounted to the forward end of the carriage.

Trend Sign Router

The Trend sign router (Illus. 34-4) copies templates on a one-to-one size ratio from a variety of letter and numeral templates available from the manufacturer, Trend Routing Technology, Watford, Hertfordshire, England. The unit can be fitted with most light-duty, hand-held routers. This Trend product will also accommodate a circle-cutting attachment.

Lobo Copy Carving Machine

The Lobo copy carving machine is designed to do three-dimensional carvings up to 6″ in diameter and 19″ long. The master copy and workpiece are held on work centers and rotated in unison. A Bosch router duplicates the cuts in the workpiece as a hand-controlled stylus follows the contour of the master pattern. Made of all-aluminum construction with a gross weight of 62 pounds, this machine will also do relief or surface carvings with a flat back. The maximum area it can carve is 12″ × 20″. Contact Lobo Power Tools, Inc., Pico Rivera, California, for more information.

Marlin Carving Machines

The Marlin carving machines (Illus. 34-5) are available in three different sizes, but all operate on the pantograph principle. They are widely used for relief flat-work carving, by professional signmakers (Illus. 34-6), and for three-dimensional carving by wood crafters, musical instrument makers, furniture-makers, etc. (Illus. 34-7 and 34-8). The carving head is any router motor unit with a 3½″ diameter. Three-dimensional carvings up to 14″ in diameter × 60″ long are possible on the largest Marlin carving machine. All models have a 3″ depth-of-cut capacity, but this depth-cutting capacity can be easily increased by adding blocks to the corners of the machine. The manufacturer also offers a series of picture templates, corporate logos, and custom logos and templates for sign- and flat-panel carving. Marlin carving machines are made by the Marlin Division of Terrco, Inc., Watertown, South Dakota.

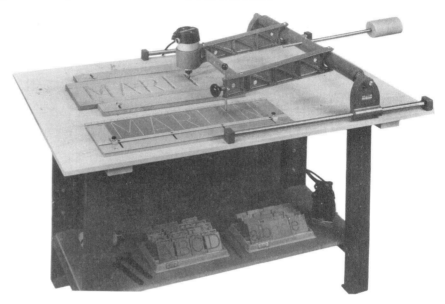

Illus. 34-5. One of three different Marlin carving machines that operate on the pantograph principle.

Illus. 34-6. A relief sign carved with a Marlin carving machine.

Illus. 34-7. A 360-degree carving duplication of an ornate furniture leg with a Marlin carving machine.

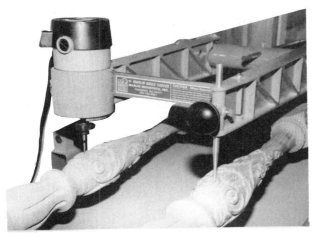

Illus. 34-8. A close-up look at the detailed, three-dimensional carving.

Dupli-Carver Machines

The Dupli-Carver machines (Illus. 34-9 and 34-10) are also router-motor-based carving machines that are available in several sizes. The largest machine, the F-200 model, can be used on workpieces as great in diameter as 14″ and as high as 40″. Illus. 34-10 shows the various carving actions possible. The machines are provided with a speed control for their one hp, standard router motor. The speed control reduces machine noise, slows the rpm, prevents overheating, and increases the tool life

of the hss bits. The Dupli-Carver machines can be used to do flat-back relief carvings, full statuary-carving (Illus. 34-11), or three-dimensional spindle-carving (Illus. 34-12). The Dupli-Carver series of machines are manufactured by Wood Miser Products and distributed by the Marlin Division of Terrco, Watertown, South Dakota.

Illus. 34-9. The Dupli-Carver machine can make three-dimensional carvings in workpieces with diameters as great as 14″ and as tall as 40″.

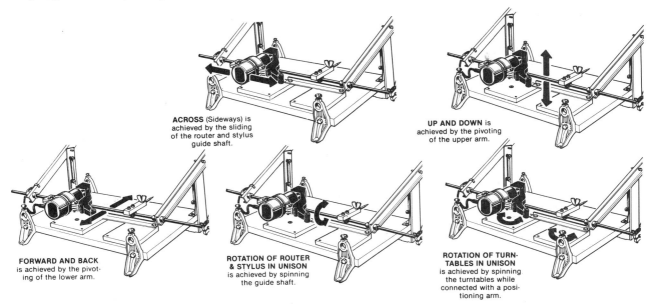

ACROSS (Sideways) is achieved by the sliding of the router and stylus guide shaft.

UP AND DOWN is achieved by the pivoting of the upper arm.

FORWARD AND BACK is achieved by the pivoting of the lower arm.

ROTATION OF ROUTER & STYLUS IN UNISON is achieved by spinning the guide shaft.

ROTATION OF TURN-TABLES IN UNISON is achieved by spinning the turntables while connected with a positioning arm.

Illus. 34-10. The five carving actions (motions) of the Dupli-Carver.

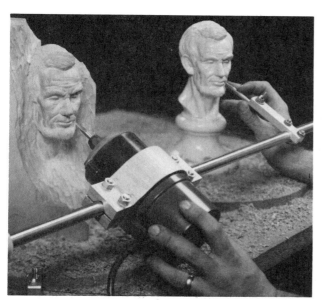

Illus. 34-11. The Dupli-Carver in use.

Illus. 34-12. Reproducing a carved table leg on a Dupli-Carver machine. Shown here is a roughing operation. Note the simple bar arrangement linking the turntables so they rotate in unison.

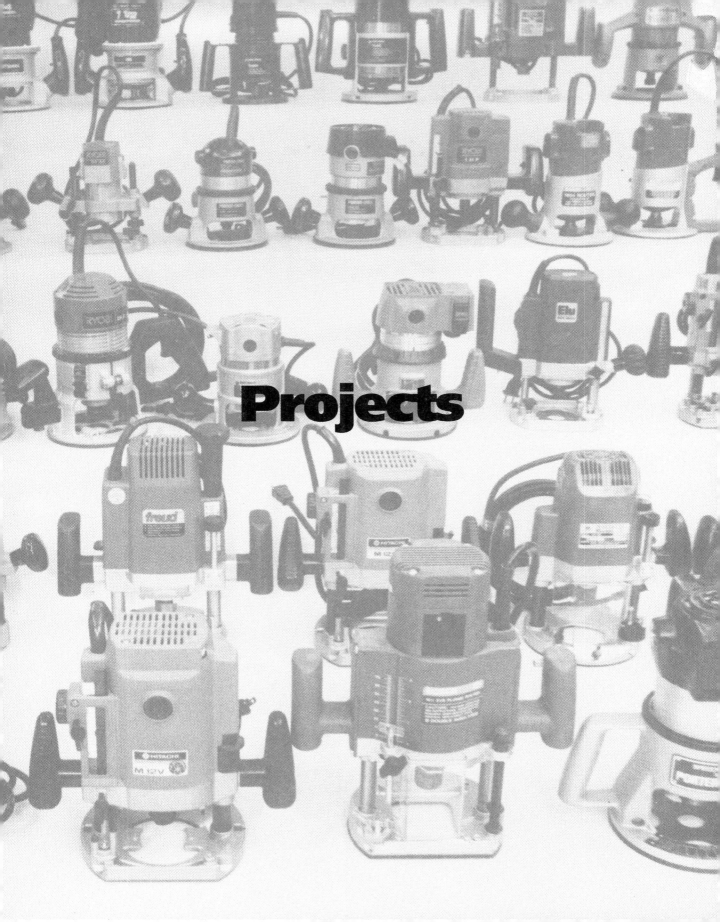

Projects

35
Projects

This chapter contains photographs and drawings showing the construction details for some of the projects displayed throughout this book. The dimensions and other specifications provided here are only suggestions. Change or modify them to accommodate the materials available or to match the capacity of your router and the limitations of your bits.

Wooden Chain

Making wooden chain (Illus. 35-1) is, without dispute, both unusual and fun. A choice of two suggested sizes is detailed in Illus. 35-2. To make the chain links, a box-type and a plug-type workpiece-holding fixture/pattern must be accurately made for each size of chain link. Make the box fixture/pattern first (Illus. 35-3). Then make the plug fixture/pattern so it fits snugly (without too much clearance) into the box pattern. Trial-and-error fitting may be necessary.

If the inside opening of the link is cut too large, the work can be shimmed with paper when set onto the plug fixture. However, for production of links, it's best to cut the inside opening of the link precisely—it makes the job faster and more fun. The majority of steps involved are shown in Illus. 35-3–35-11.

Illus. 35-1. Chain can be made in large or small sizes.

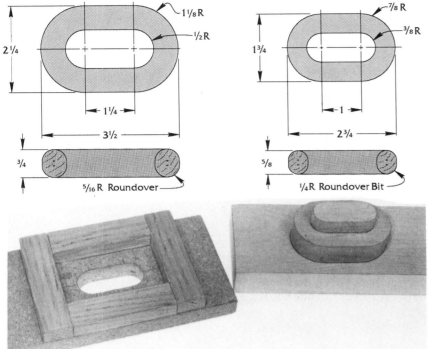

Illus. 35-2. Details for making chain in two sizes.

Illus. 35-3. Make work-holding fixtures as shown to match the chain-link size. The box fixture is shown on the left.

Illus. 35-4. Prepare the blocks to their finished sizes. Remove most of the inside waste by boring. Use a ball-bearing trimming bit to clean up the inside opening of the plug fixture with it held in the box fixture.

Illus. 35-5. Here's how the inside opening is cleaned and trimmed to size on the box fixture. Note that the ball bearing rides on the pattern part of the fixture.

Illus. 35-6. The outside profile can be shaped on the plug fixture with a trimming bit. Feed the fixture counterclockwise. To prevent chipping, cut only the two opposite corners, as shown. Then flip the workpiece over and cut the two remaining corners. This procedure allows you to cut all the corners working with the grain. A piloted spiral bit works best for this job. Another option is to saw the corners to rough size and then trim them using the setup shown here.

Illus. 35-7. Rounding over the inside edges with the work held in the box fixture.

Illus. 35-8. Rounding over the outside edges on the plug fixture. Set the depth of cut so a slight flat remains for the ball bearing to ride on.

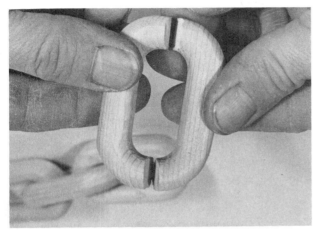

Illus. 35-9. After sanding, split every other link by hand. If they are difficult to snap, use a thin knife and open one end by cutting from the inside.

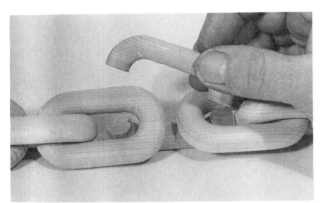

Illus. 35-10. The snapped links will glue back together perfectly to make a length of chain, as shown.

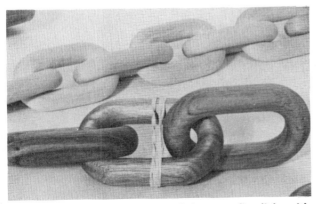

Illus. 35-11. First, assemble sets of threes or five links with rubber-band pressure, as shown. Then connect the sets using clamps, tape, or hand pressure.

Miscellaneous Projects

Illus. 35-12—35-17 are photographs and patterns depicting picture-frame, hexagon hanging-lamp, and cutting-board projects.

Illus. 35-12. This double-oval picture frame requires freehand incise-routing. Cut rabbets on the inside edges of the ovals and round over all forward edges.

Illus. 35-13. Pattern for the picture-frame project.

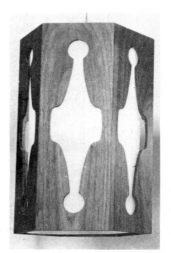 

Illus. 35-14 (right). Hexagon hanging-lamps made of ¼″ ply-wood. Thin fibreglass shade material rolled into cylinders for the inside is available from craft stores and mail-order houses specializing in plastic material. Illus. 35-15 (below). Half-pattern details for the hanging lamp.

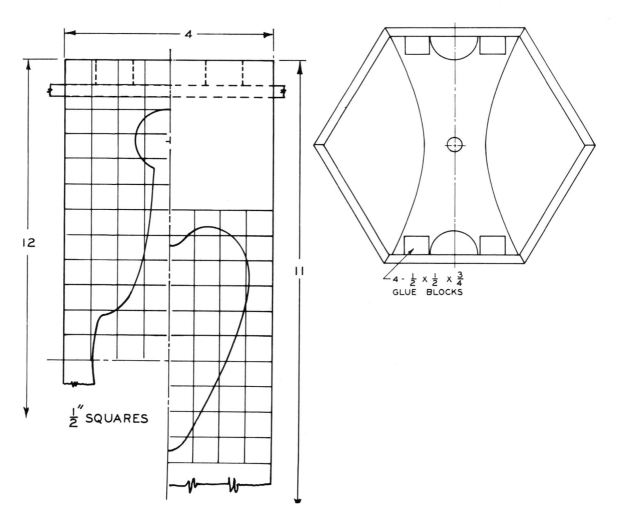

Illus. 35-16. This cutting board with a slanted juice-drainboard surface is a good project for template routing.

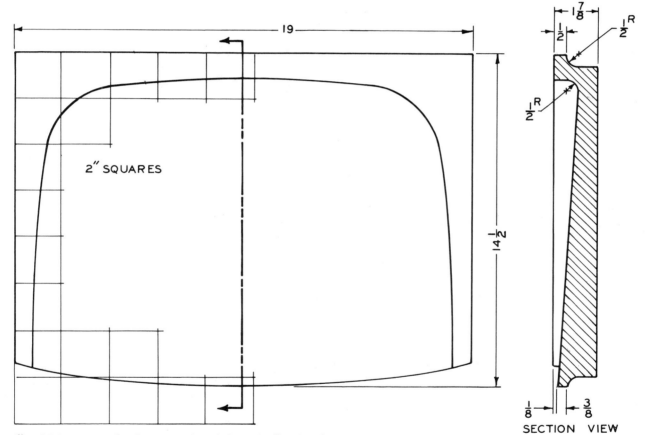

Illus. 35-17. Pattern for the cutting board shown in Illus. 35-16.

Metric Equivalents

INCHES TO MILLIMETRES AND CENTIMETRES

MM—millimetres　　*CM—centimetres*

Inches	MM	CM	Inches	CM	Inches	CM
⅛	3	0.3	9	22.9	30	76.2
¼	6	0.6	10	25.4	31	78.7
⅜	10	1.0	11	27.9	32	81.3
½	13	1.3	12	30.5	33	83.8
⅝	16	1.6	13	33.0	34	86.4
¾	19	1.9	14	35.6	35	88.9
⅞	22	2.2	15	38.1	36	91.4
1	25	2.5	16	40.6	37	94.0
1¼	32	3.2	17	43.2	38	96.5
1½	38	3.8	18	45.7	39	99.1
1¾	44	4.4	19	48.3	40	101.6
2	51	5.1	20	50.8	41	104.1
2½	64	6.4	21	53.3	42	106.7
2	76	7.6	22	55.9	43	109.2
3½	89	8.9	23	58.4	44	111.8
4	102	10.2	24	61.0	45	114.3
4½	114	11.4	25	63.5	46	116.8
5	127	12.7	26	66.0	47	119.4
6	152	15.2	27	68.6	48	121.9
7	178	17.8	28	71.1	49	124.5
8	203	20.3	29	73.7	50	127.0

Current Books by Patrick Spielman

Carving Wild Animals: Life-Size Wood Figures. Spielman and renowned woodcarver Bill Dehos show how to carve more than 20 magnificent creatures of the North American wild. A cougar, black bear, prairie dog, squirrel, raccoon, and fox are some of the life-size animals included. Step-by-step, photo-filled instructions and multiple-view patterns, plus tips on the use of tools, wood selection, finishing, and polishing, help bring each animal to life. Oversized. Over 300 photos. 16 pages in full color. 240 pages.

Christmas Scroll Saw Patterns. Patrick and Patricia Spielman provide over 200 original, full-size scroll saw patterns with Christmas as the theme, including: toys, shelves, tree, window, and table decorations; segmented projects; and alphabets. A wide variety of Santas, trees, and holiday animals are included, as is a short, illustrated review of scroll saw techniques. 4 pages in color. 164 pages.

Classic Fretwork Scroll Saw Patterns. Spielman and coauthor James Reidle provide over 140 imaginative patterns inspired by and derived from mid- to late-19th-century scroll-saw masters. This book covers nearly 30 categories of patterns and includes a brief review of scroll-saw techniques and how to work with patterns. The patterns include ornamental numbers and letters, beautiful birds, signs, wall pockets, silhouettes, a sleigh, jewelry boxes, toy furniture, and more. 192 pages.

Country Mailboxes. Spielman and coauthor Paul Meisel have come up with the 20 best country-style mailbox designs. They include an old pump fire wagon, a Western saddle, a Dalmatian, and even a boy fishing. Simple instructions cover cutting, painting, decorating, and installation. Over 200 illustrations. 4 pages in color. 164 pages.

Gluing & Clamping. A thorough, up-to-date examination of one of the most critical steps in woodworking. Spielman explores the features of every type of glue—from traditional animal-hide glues to the newest epoxies—the clamps and tools needed, the bonding properties of different wood species, safety tips, and all techniques from edge-to-edge and end-to-end gluing to applying plastic laminates. Also included is a glossary of terms. Over 500 illustrations. 256 pages.

Making Country-Rustic Wood Projects. Hundreds of photos, patterns, and detailed scaled drawings reveal construction methods, woodworking techniques, and Spielman's professional secrets for making indoor and outdoor furniture in the distinctly attractive Country-Rustic style. Covered are all aspects of furniture making from choosing the best wood for the job to texturing smooth boards. Among the dozens of projects are mailboxes, cabinets, shelves, coffee tables, weather vanes, doors, panelling, plant stands, and many other durable and economical pieces. 400 illustrations. 4 pages in color. 164 pages.

Making Wood Bowls with a Router & Scroll Saw. Using scroll-sawn rings, inlays, fretted edges, and much more, Spielman and master craftsman Carl Roehl have developed a completely new approach to creating decorative bowls. Over 200 illustrations. 8 pages in color. 168 pages.

Making Wood Decoys. This clear, step-by-step approach to the basics of decoy carving is abundantly illustrated with close-up photos for designing, selecting, and obtaining woods; tools; feather detailing; painting; and finishing of decorative and working decoys. Six different professional decoy artists are featured. Photo gallery (4 pages in full color) along with numerous detailed plans for various popular decoys. 164 pages.

Making Wood Signs. Designing, selecting woods and tools, and every process through finishing clearly covered. Instructions for hand- and power-carving, routing, and sandblasting techniques for small to huge signs. Foolproof guides for professional letters and ornaments. Hundreds of photos (4 pages in full color). Lists sources for supplies and special tooling. 148 pages.

New Router Handbook. This updated and expanded

version of the definitive guide to routing continues to revolutionize router use. The text, with over 1,000 illustrations, covers familiar and new routers, bits, accessories, and tables available today; complete maintenance and safety techniques; a multitude of techniques for both hand-held and mounted routers; plus dozens of helpful shop-made fixtures and jigs. 384 pages.

Original Scroll Saw Shelf Patterns. Patrick Spielman and Loren Raty provide over 50 original, full-size patterns for wall shelves, which may be copied applied directly to wood. Photographs of finished shelves are included, as well as in formation on choosing woods, stack sawing, and finishing. 4 pages in color. 132 pages.

Realistic Decoys. Spielman and master carver Keith Bridenhagen reveal their successful techniques for carving, feather texturing, painting, and finishing wood decoys. Details you can't find elsewhere—anatomy, attitudes, markings, and the easy, step-by-step approach to perfect delicate procedures—make this book invaluable. Includes listings for contests, shows, and sources of tools and supplies. 274 close-up photos. 8 pages in color. 232 pages.

Router Basics. With over 200 close-up, step-by-step photos and drawings, this valuable starter handbook will guide the new owner, as well as provide a spark to owners for whom the router isn't the tool they turn to most often. Covers all the basic router styles, along with how-it-works descriptions of all its major features. Includes sections on bits and accessories, as well as square-cutting and trimming, case and furniture routing, cutting circles and arcs, template and freehand routing, and using the router with a router table. 128 pages.

Router Jigs & Techniques. A practical encyclopedia of information, covering the latest equipment to use with the router, it describes all the newest commercial routing machines, along with jigs, bits, and other aids and devices. The book not only provides invaluable tips on how to determine which router and bits to buy, it explains how to get the most out of the equipment once it is bought. Over 800 photos and illustrations. 384 pages.

Scroll Saw Basics. Features more than 275 illustrations covering basic techniques and accessories. Sections include types of saws, features, selection of blades, safety, and how to use patterns. Half a dozen patterns are included to help the scroll saw user get started. Basic cutting techniques are covered, including inside cuts, bevel cuts, stack-sawing, and others. 128 pages.

Scroll Saw Country Patterns. With 300 full-size patterns in 28 categories, this selection of projects covers an extraordinary range, with instructions every step of the way. Projects include farm animals, people, birds, and butterflies, plus letter and key holders, coasters, switch plates, country hearts, and more. Directions for piercing, drilling, sanding, and finishing, as well as tips on using special tools. 4 pages in color. 196 pages.

Scroll Saw Fretwork Patterns. This companion book to *Scroll Saw Fretwork Techniques & Projects* features over 200 fabulous, full-size fretwork patterns. These patterns include popular classic designs, plus an array of imaginative contemporary ones. Choose from a variety of numbers, signs, brackets, animals, miniatures, and silhouettes, and more. 256 pages.

Scroll Saw Fretwork Techniques & Projects. A study in the historical development of fretwork, as well as the tools, techniques, materials, and project styles that have evolved over the past 130 years. Every intricate turn and cut is explained, with over 550 step-by-step photos and illustrations. 32 projects are shown in full color. The book also covers some modern scroll sawing machines as well as state-of-the-art fretwork and fine scroll-sawing techniques. 8 pages in color. 232 pages.

Scroll Saw Handbook. The workshop manual to this versatile tool includes the basics (how scroll saws work, blades to use, etc.) and the advantages and disadvantages of the general types and specific brand-name models on the market. All cutting techniques are detailed, including compound and bevel sawing, making inlays, reliefs, and recesses, cutting metals and other non-woods, and marquetry. There's even a section on transferring patterns to wood. Over 500 illustrations. 256 pages.

Scroll Saw Holiday Patterns. Patrick and Patricia Spielman provide over 100 full-size, shaded patterns for easy cutting, plus full-color photos of projects. Will serve all your holiday pleasures—all year long. Use these holiday patterns to create decorations, centerpieces, mailboxes, and diverse projects to keep or give as gifts. Standard holidays, as well as the four seasons, birthdays, and anniversaries, are represented. 8 pages of color. 168 pages.

Scroll Saw Pattern Book. The original classic pattern book—over 450 patterns for wall plaques, refrigerator magnets, candle holders, pegboards, jewelry, ornaments, shelves, brackets, picture frames, signboards, and many other projects. Beginning and experienced scroll saw users alike will find something to intrigue and challenge them. 256 pages.

Scroll Saw Patterns for the Country Home. Patrick and Patricia Spielman and Sherri Spielman Valitchka produce a wide-ranging collection of over 200 patterns on country themes, including simple cutouts, mobiles, shelves, sculpture, pull toys, door and window toppers, clock holders, photo frames, layered pictures, and more. Over 80 black-and-white photos and 8 pages of color photos help you to visualize the steps involved as well as the finished projects. General instructions in Spielman's clear and concise style are included. 200 pages.

Scroll Saw Puzzle Patterns. 80 full-size patterns for jigsaw puzzles, stand-up puzzles, and inlay puzzles. With meticulous attention to detail, Patrick and Patricia Spielman provide instructions and step-by-step photos, along with tips on tools and wood selection, for making dinosaurs, camels, hippopotami, alligators—even a family of elephants! Inlay puzzle patterns include basic shapes, numbers, an accurate piece-together map of the United States, and a host of other colorful educational and enjoyable games for children. 8 pages of color. 264 pages.

Scroll Saw Shelf Patterns. Spielman and Loren Raty offer full-size patterns for 44 different shelf styles. Designs include wall shelves, corner shelves, and multi-tiered shelves. The patterns work well with 1/4-inch hardwood, plywood or any solid wood. Over 150 illustrations. 4 pages in color. 132 pages.

Scroll Saw Silhouette Patterns. With over 120 designs, Spielman and James Reidle provide an extremely diverse collection of intricate silhouette patterns, ranging from Victorian themes to sports to cowboys. They also include mammals, birds, country and nautical designs, as well as dragons, cars, and Christmas themes. Tips, hints, and advice are included along with detailed photos of finished works. 160 pages.

Sharpening Basics. The ultimate handbook that goes well beyond the "basics," to become the major up-to-date reference work features more than 300 detailed illustrations (mostly photos) explaining every facet of tool sharpening. Sections include bench-sharpening tools, sharpening machines, and safety. Chapters cover cleaning tools, and sharpening all sorts of tools, including chisels, plane blades (irons), hand knives, carving tools, turning tools, drill and boring tools, router and shaper tools, jointer and planer knives, screwdrivers and scrapers, and, of course, saws. 128 pages.

Spielman's Original Scroll Saw Patterns. 262 full-size patterns that don't appear elsewhere feature teddy bears, dinosaurs, sports figures, dancers, cowboy cutouts, Christmas ornaments, and dozens more. Fretwork patterns are included for a Viking ship, framed cutouts, wall-hangers, key-chain miniatures, jewelry, and much more. Hundreds of step-by-step photos and drawings show how to turn, repeat, and crop each design for thousands of variations. 4 pages of color. 228 pages.

Victorian Gingerbread: Patterns & Techniques. Authentic pattern designs (many full-size) cover the full range of indoor and outdoor detailing: brackets, corbels, shelves, grilles, spandrels, balusters, running trim, headers, valances, gable ornaments, screen doors, pickets, trellises, and much more. Also included are complete plans for Victorian mailboxes, house numbers, signs, and more. With clear instructions and helpful drawings, the book also provides tips for making gingerbread trim. 8 pages in color. 200 pages.

Victorian Scroll Saw Patterns. Intricate original designs plus classics from the 19th century are presented in full-size, shaded patterns. Instructions are provided with drawings and photos. Projects include alphabets and numbers, silhouettes and designs for shelves, frames, filigree baskets, plant holders, decorative boxes, picture frames, welcome signs, architectural ornaments, and much more. 160 pages.

Working Green Wood with PEG. Covers every process for making beautiful, inexpensive projects from green wood without cracking, splitting, or warping it. Hundreds of clear photos and drawings show every step from obtaining the raw wood through shaping, treating, and finishing PEG-treated projects. 175 unusual project ideas. Lists supply sources. 120 pages.

Photo Credits

The illustrations in this book display the products, creations, and photography of many people and business organizations. Represented among them are the following: Albert Constantine and Son, Inc., Illus. 14-9, 14-10, and 14-11; Bonville, Keith, 13-39 and 35-16; Black & Decker, 16-6, 8-19, 9-41, 9-42, 10-9, 10-21, and 16-6; Buckeye Tools Corp., 2-58; Bosch, 5-33; Cascade Tools, 7-7; Ekstrom, Carlson & Co., 5-4, 5-5, and 5-30; E.L. Bruce Co., 11-27 (photo by Hedrich-Blessing); Forest City Tool Co., 5-80; Hartlauer, Walter, 18-14; Porter-Cable, i-10, 5-3, 5-57, 5-58; Scan Furnishing, 11-51; Sears, 18-11, 18-12, 18-13, 33-1, and 33-2; Shelburne Museum, Inc., 1-10 (photo by Einers J. Mengis); Shopsmith, Inc., 30-21.

Index